Praise for

THE SUN IS A COMPASS

"Imagine trekking four times farther than Cheryl Strayed did in *Wild*, without a trail, through swarms of mosquitoes thick enough to suck caribou dry. In this marvelous tale of grit and grace, ornithologist Caroline Van Hemert leaves behind a lab full of caged chickadees to embark on her own epic migration to the Arctic, reconnecting with the reverence for nature that drove her to science in the first place. For those of us less skilled at fashioning our own rowboats, dodging avalanches, and fending off hungry bears, this intimate book is a precious window into a remote wilderness of formidable beauty."

—Emily Voigt, author of *The Dragon Behind the Glass*

"Ornithologist and naturalist Caroline Van Hemert has written a thrilling account of her journey from the Pacific Coast to the Arctic Ocean. A triumph in wilderness travel, scientific curiosity, and adventure writing that exposes the sublime thrill and loving touch to be found in nature and our fellow human beings."

—John Marzluff, professor of wildlife science and author of *Welcome to Subirdia*

"Most adventures start with a map, often following a route that looks possible on paper but turns out to be less than possible in real life. Caroline Van Hemert's *The Sun Is a Compass* tells of a journey that looked almost impossible on a map. I was left with the same sense of amazement I have felt, five miles from land in the Gulf of Alaska, with nothing in bloom for a thousand miles, when a hummingbird flies by, circles once, and continues north."

—George Dyson, author of *Turing's Cathedral*

"*The Sun Is a Compass* is an adventure story but also a love story. It is thrilling, uplifting, and hopeful, both as a journey across northern wilds and as a diary of a couple growing ever closer together. Caroline and Pat's journey will rekindle your faith in human endurance, and intimacy."

—David Rothenberg, author of *Nightingales in Berlin* and *Why Birds Sing*

"Caroline Van Hemert has crafted a book as remarkable and dimensional as her epic journey. She is able to offer a scientist's insight into the natural world while writing of danger, beauty, and love without ego and with refreshing grace and honesty. Her book is a gift not just to those who like to venture on the wild side, but to anyone intrigued by the possibilities of strong partnership, imagination, and curiosity. This is unlikely to be a book you just read; it is one that will make you soar."

—Jill Fredston, author of *Rowing to Latitude* and *Snowstruck*

"In *The Sun Is a Compass,* adventure and romance journey hand in hand, covering 4,000 tough miles, reminding all of us that the easy way may not be the best way."

—Bill Streever, author of *Cold*

"In a time when stories of extreme outdoor adventures have become commonplace, Caroline Van Hemert's *The Sun Is a Compass* stands out because it is at heart a love story. A remarkably skilled and experienced wilderness traveler, the author writes in the clear language of a scientist who observes her world through the eyes of a poet, across 4,000 miles of risk and endurance, in concert with an extraordinary man. It's a hell of a read."

—Lynn Schooler, author of *Walking Home*

THE SUN IS
A COMPASS

A 4,000-Mile Journey into the Alaskan Wilds

Caroline Van Hemert

Little, Brown Spark
New York Boston London

Little, Brown Spark
Hachette Book Group
1290 Avenue of the Americas, New York, NY 10104
littlebrownspark.com

First Edition: March 2019

Little, Brown Spark is an imprint of Little, Brown and Company, a division of Hachette Book Group, Inc. The Little, Brown Spark name and logo are trademarks of Hachette Book Group, Inc.

The publisher is not responsible for websites (or their content) that are not owned by the publisher.

The Hachette Speakers Bureau provides a wide range of authors for speaking events. To find out more, go to hachettespeakersbureau.com or call (866) 376-6591.

Photographs by Caroline Van Hemert and Patrick Farrell
Illustrations by Patrick Farrell
Map by Colin Shanley

ISBN 978-0-316-41442-5
LCCN 2018945473

10 9 8 7 6 5 4 3 2 1

LSC-C

Printed in the United States of America

For Pat, my partner in everything

CONTENTS

CONTENTS

THE SUN IS
A COMPASS

Prologue

SWIMMING THE CHANDALAR

I'm standing on the bank of the swift Chandalar River in the Brooks Range of northern Alaska, trying to gather the courage to swim across. My husband, Pat, is by my side. We're alone, as we have been for most of the past five months.

The sky is a depthless sort of overcast, no definition in the clouds, no glimmer of sunshine. The temperature hovers just above freezing and the air is damp after a night of rain. I grip the straps of my pack, my fingers raw from the chill, and lean against Pat as we look down at the river that flows in a wide channel sixty feet below us. The only sound is the steady rush of moving water. I push away the voice in my head that echoes a single question. *What are we doing?*

It's the fifth of August, 2012. Over the last 139 days, we have traversed nearly three thousand miles, most recently through places so lightly traveled our topographic maps have little to say about them. Only the highest peaks are labeled, and then solely by elevation. The Brooks Range is the northernmost major mountain range on earth and has retained its integrity in ways that few places have. Many of the creeks and valleys are nameless, their curves and riffles left unexplored.

There are no soft edges here, no boardwalks or trails or park rangers. It's wild, empty, and gritty.

We're here because we're attempting to travel entirely under our own power from the Pacific Northwest to a remote corner of the Alaskan Arctic. We're here because we need wilderness like we need water or air. Like we need each other. For me, this trip is also a journey back to trees and birdsong, to lichen and hoof prints. Before leaving, I had lost my way on the path that carried me from biology to natural wonder. I had forgotten what it meant, not only in my mind, but in my heart, to be a scientist.

We have a thousand miles ahead of us, but for now all that matters is this river. On the map it looked harmless, squiggly and blue. As I stare down at it now, it's the color of mud. From our elevated vantage, the water's opaque surface appears smooth, but when Pat throws a spruce bough from the bank, it bobs in the small waves, spins once, then vanishes quickly downriver.

In the first months of the journey, our destination was so distant that it seemed almost peripheral. Kotzebue. A small village on the shores of the Chukchi Sea. A place on the map as arbitrary as any other. We were consumed by each day, distracted by aching muscles and whales and the simple act of moving. Always moving. But the stakes quietly grew, shape-shifting from a tally of miles into something much more. Only now am I beginning to see this trip for what it is. A celebration and a letting go of youth. A reawakening of the biologist in me. A reckoning between us and the land. Something we must see through to the end.

And so crossing this river has become necessary, in the way that it's necessary to kiss a lover before leaving, to pause and look up when the moon is rising. Our bodies know what is essential and what is not.

PROLOGUE

* * *

Before we left, people asked us why we were taking this trip; they wondered what compelled us to want to "disappear" for a while. I tried to explain that escapism wasn't our goal— neither of us was running from a broken marriage or drug addiction or academic failure. We weren't trying to set a record or achieve a first. We were simply trying to find our way home.

Shortly after Pat and I met in 2001, we discovered that we were most fully ourselves in wild places. That our love was strongest among rocks and rivers, trees and tundra. Since our first summer together, when we spent two months camped on the bank of a remote Arctic river, we had dreamed about another grand adventure. Increasingly, though, time in the outdoors was taking a backseat to more mundane endeavors. Our trips were shrinking, our commitments growing. Even worse, I had just finished a Ph.D. in biology feeling more distant than ever from the natural world. Five years of study had started as an act of love and turned into pure drudgery.

My research focused on a strange cluster of beak deformities that had recently emerged among Alaskan chickadees and other birds. The afflicted birds grew curled and grotesque beaks that resembled something from a dark version of Dr. Seuss. When I began my graduate project, I was sure I could find answers to the mystery of the beak deformities, and that the resulting facts would matter. I fancied myself something of a wildlife detective, searching for clues that would help me crack the case. But instead I quickly learned that the most basic information about the anatomy of a bird's beak was not yet available, and I had no choice but to ask the simplest questions first. I began with the tedious, unglamorous work of slicing beaks into impossibly thin pieces using miniature knives and examining them under high-powered microscopes. I housed chickadees in a laboratory and studied the way their beaks grew, feeling remorse each time I stepped into the room and stared at two dozen pairs of eyes that would never again see birch leaves fluttering in the wind or probe a tree's bark for spiders and beetles.

The tiny black-capped chickadees whose familiar calls belie the fact that they are actually one of the most remarkable species on earth were first my inspiration and then, later, my bane. When my advisor toasted me after my dissertation defense, I cringed, knowing I had failed in the most fundamental of ways. This wasn't a failure in the traditional sense—my calculations stood up to scrutiny, my experiments worked, my chapters were well written. But underneath it all was the ugly fact that I simply didn't care anymore. Between hundreds of hours peering under a microscope and observing chickadees in cages, I had forgotten why I'd wanted to be a biologist in the first place.

During the years of my graduate research, Pat dedicated himself to several building projects and spent more time communing with hammer and saw than with forests or mountains. Since he was a boy, he had been driven to build things. His elaborate childhood forts eventually gave way to cabins and houses, and he had created a fledgling, but successful, design-and-build company. But he was tired of managing budgets, juggling material orders, and shoring up leaky foundations. He questioned why he wasted sunny afternoons buried in drywall dust only to realize that building houses, even those he designed, would never be enough.

In our commitment to education and jobs, we had neglected what mattered most to us. Our calendars were shaped by academic deadlines and construction schedules rather than tide cycles and seasons. We missed the freedom that came with sleeping outdoors for weeks or months at a time. Recently, decisions about whether to have children and how to care for aging parents had started to feel pressing. My dad had been diagnosed with a degenerative neurological disease. My younger sister was pregnant. The career that awaited me felt increasingly like a sentence rather than an opportunity. Still, I wasn't entirely sure what all of this had to do with our trip or what I hoped to find along the way. I didn't yet understand how traveling across four thousand miles of wilderness would help me face my looming adulthood or a job I wasn't sure I wanted. I didn't realize I needed to find my way back to biology by the same means I had first discovered it.

Only months after we left did I begin to appreciate that this trip offered what ordinary life could not. Clear edges. Truth. Acceptance. An understanding that living with uncertainty is not only OK; it is the only option. Before we started, I wanted nothing to do with the facts that were staring back at me. Life is tenuous. Love is risky. We have so much to lose along the way. I had forgotten the converse side of this equation, that the most precious things in life are those that don't last forever. I needed a crash course outdoors to remind myself that a life is not merely a tally of days, that what really matters cannot be quantified. The glimpse of a wolf's tawny back, his coat shimmering with dew. The sound of my dad's voice on the satellite phone, holding steady and sure. The look Pat gives me when he knows my pack straps are cutting into my shoulders and my spirit is waning, his expression encouraging me that I *can* do the impossible.

We hadn't originally planned to swim across anything. But now we're perched at the edge of a cold Arctic river without our packrafts; they are on a mail plane heading west. Several days ago, we decided we would shed the extra weight of our boats to lighten our loads. We'll pick them up again in the village of Anaktuvuk Pass, two hundred miles to the west, after much of the steep terrain is behind us. Winter is only weeks away and we need to move quickly if we're to reach Kotzebue before freeze-up. The first season's snow fell last week as we woke to caribou milling around our tent, a small band traveling south. Before we got to the river, our decision to ship the boats seemed like a good one. Now, I'm not so sure.

As we watch the water swirling below, I try to guess how long it might take to swim to the other side, two hundred yards away. Five minutes? Ten? Just as I realize that the distance is equivalent to several laps in a very cold pool, Pat interrupts my calculations by asking where I think we should cross. Before the bend or after? Where the river is widest or narrowest? I see him looking downriver. He is thinking the same thing I am. *Where will we end up if we get carried downstream?*

We climb down the bank and find a spot to enter, right past a large elbow in the river, and I empty the contents of my pack, searching for the thin waterproof bag that contains my extra clothes. Out comes my sleeping bag, sleeping pad, three stuff sacks of food, satellite phone, rain gear, cooking pot, and camera. When I find the clothes, I begin to undress, goose bumps rising on my skin in the cool air. I re-layer with almost everything I'm carrying—wool long-underwear tops and bottoms, fleece pullover, synthetic vest, nylon pants, and a wool hat—and wiggle into a plastic trash bag with holes cut for my head and arms before pulling on my rain jacket and pants. We know the water will penetrate our layers, but are hoping the rain gear and plastic bag will help to preserve our body heat against the cold. *Like an improvised wet suit,* Pat explained when he came up with the idea.

I volunteer to go first, not because I'm feeling especially

brave, but because one of us must do it. Pat isn't one for chauvinism; still, he hesitates for several moments, staring across the water. He only agrees when I explain that it will be easier for him to rescue me than the reverse. *Just in case,* I add.

Late last night, curled in our sleeping bags, I tried to envision our crossing. I told Pat that if it seemed too dangerous to swim maybe we could hike back to the nearest village and find someone to give us a ride to the other side. But even as I said this, both of us knew it wouldn't happen. It would mean we had failed.

When we committed to this project—to travel from rainforest to ice-filled sea, from the edge of the continental United States to the edge of the earth—we decided it would be completely on our terms. No roads, no trails, and no motors. We would travel by foot, on skis, in rowboats, rafts, and canoes. We would use only our own muscles to carry us through some of the wildest places left on earth. This wasn't a mandate borne purely of stubbornness, though Pat and I each possess a healthy dose of that trait, but because it would allow us to know the landscape as intimately as we knew each other. Just getting to remote places wasn't the point. We could have hired a plane to drop us off at any number of locations that would qualify as the middle of nowhere. But we wanted something different. We wanted to hear the crunch of lichen beneath our feet, to smell the tundra after a rainstorm, to understand how it felt to walk in a caribou's tracks or paddle alongside a beluga whale.

For years, adventure was simply a part of our lives. It hadn't yet taken on the urgency that arrived, in my early thirties, like a loud and obtrusive neighbor, as my perception of time shifted from lazy and boundless to precious and finite.

With it came the understanding that youth is only a temporary pause, a whistle-stop on the train that barrels along, leaving the aging and frail and ill—in the end, all of us—behind. When we first started planning, I had an inkling that this trip would matter more than all of the others we'd taken in our ten years together. Not just because of its scale, which was quickly growing to outrageous proportions, but because if we didn't do it now, we might never have another chance. We knew our bodies wouldn't stay strong forever. Inevitably, our responsibilities would grow; our freedom would shrink. I would never again be a thirty-three-year-old on the brink of finishing her Ph.D., childless, disillusioned by the prospect of an academic career, and convinced that whatever it was I needed could be found between two distant places on the map, one a coastal town where I had met my husband, the other a remote, ice-locked land I'd never seen.

I'm shivering before I step into the river. When I begin to wade, the mud soft and forgiving beneath my feet, icy water seeps quickly up my pant legs. My muscles stiffen in response, my knees suddenly wooden, my groin aching. Several steps later I lose contact with the bottom as the current tugs on my hips.

Immediately, I'm being carried downstream, farther from Pat but no closer to the other side. I need to start swimming, and fast. I lace my arms backward through the straps of my pack and attempt to balance my chest on top of the buoyant load as though it is a kickboard. For a moment, this seems to be working. I'm floating and kicking. But my upper body is perched so high above the surface that I can't get any purchase with my flailing legs.

I try again. Lowering my body and leveraging my chin against the bottom of my pack, I kick like hell. I can barely see above the pack, and when I crane my neck, breathing hard, I realize I'm paralleling the shore. I reorient myself and try once more. I flutter my feet but nothing happens. I kick from my hips, but I only move farther downstream. *This isn't working. Hurry up.*

As I'm floundering, I think of my mom, queen of the breaststroke. *Frog kicks? Maybe?* After my first contorted attempts, I find a way to use not just my legs but my arms, sliding abbreviated strokes through the shoulder straps. I direct my pack with my chin. It works. I can move *and* steer and begin to propel myself toward the middle of the channel. Soon, Pat yells from the bluff above that I've made it halfway.

I cheer myself on silently, focusing the only part of my gaze that isn't blocked by my pack onto the trees that are growing larger with each stroke. I can see my progress. *Better. Almost there.* A surge of confidence follows, and I slow my frantic motions enough to catch my breath. Seconds later, I hit a stiff eddy line. A dozen yards from shore, the swirling water leaves me nearly stationary. Pat shouts something unintelligible. I try to stand up, but a small creek joins the river here and the water is surprisingly deep.

Pat yells again. This time, I hear "Get up!"—but I can't. I'm suddenly afraid. And starting to tire. My inner voice wavers. *If you stop now. You. Will. Wash. Away. Act, don't think, Caroline.* I force my mind to go still. Robotic. *Kick hard. Harder.* I try to touch down again, but feel only water beneath my feet. I close my eyes and channel everything into my legs. *Do it. Or else.*

After several more attempts, I feel a release. I have finally managed to break through the eddy. As soon as I find contact with the muddy bottom, I wade out of the water and flop onto the shore. I take several breaths lying down, staring up at the sky. When I raise my head and look across the river,

I see Pat pump his fist into the air, celebrating for me. I'm only partially relieved. The swim was much worse than I had imagined. Now I have to watch Pat take a turn. He's a strong swimmer, but the river's stronger.

As I stand up and move away from the river's edge, Pat finishes stuffing the last items into his pack. It takes forever. He seals his pack, then opens it up again, retrieving something he left behind on the ground. He arranges and rearranges his load, my anxiety building with each adjustment. When he finally scrambles down the cutbank, he looks small and the river huge.

Within seconds of wading into the water, he's kicking his legs and windmilling his right arm, holding the pack with his left. But I'm not sure his one-armed crawl is working. All but the top of his head is obscured by splashing. Partway across, he switches arms. He slows for a moment and begins to drift downstream. "Come on, Pat," I yell, willing away the excruciating minutes of watching him struggle, and he begins to windmill again. When he's finally near enough for me to see his face, his expression terrifies me.

He's wide-eyed and intense. Fighting. Hard.

"Are you OK?" I shout. No response. He hesitates and changes arms. I shout again. Nothing. Fifty yards from shore he's practically at a standstill. I scream that if he doesn't answer me I'm coming in after him.

"Hold on to your pack, I'll be there in a second!" Still no answer. He's moving toward me so slowly he looks stationary. I wade into the water and begin to breaststroke through the eddy, cursing myself for waiting so long. If the current carries Pat much farther, I might not be able to reach him in time. And even if I do, I'm not sure I can help.

I barely notice the cold this time as I pull against the gray water. Beneath the surface the current churns and grasps.

Even without my backpack, it takes all of my energy to fight through the eddy again. Pat stares intently at the shore and mumbles that he is tired, so tired. Fatigue is only part of the problem. Nearly ten minutes in the frigid river is long enough for hypothermia to set in. When I'm close enough to touch him, I grab his pack and position myself behind him. Without the pack, he can use both of his arms and paddles more smoothly. At the eddy, he glances back at me before stroking hard for shore. I'm right behind him, harnessing the strength that comes with fear. Finally, we stumble out of the water and collapse together on the riverbank.

Horror at what could have just happened replaces the adrenaline coursing through me.

"Damn," Pat says and shakes his head, his eyes shining against the leaden sky. He shivers as he explains that his jacket had filled with water, making it difficult to lift his arm with each stroke. Suddenly, I understand exactly how much I stand to lose. Underlying all of our choices is the fact that if something happened to one of us, the other would have to face the consequences. At times like these, it's impossible not to question whether the risk is worth the reward. Whether we are asking too much of the land, and of ourselves. We stand up, hug each other tightly, and begin to strip off our sodden clothes. Pat jumps up and down to warm himself. I help him with the zippers of his jacket, then work on my own layers.

As I'm wringing out my shirt, contemplating what we're doing here, I hear a sound I've never heard before. I pause, grab Pat's arm, and put my finger to my lips. *Listen,* I whisper. Silence. And then I hear it again. A familiar *chick-a-dee-dee-dee.* But from behind the voice emerges something entirely new. Coarser, more nasal, perhaps an extra scolding *tisk* at the beginning. The differences are subtle, and I strain to hear each note.

"Oh my god, Pat. I think it's a gray-headed chickadee!"

A moment later, I see not just one bird but an entire family of chickadees flutter onto a nearby spruce tree. Perched on a branch, watching us, are two adults and four fuzzy young.

A gray-headed chickadee is anything but glamorous. As the name describes, it's gray. And small. And very, very hard to find. So hard, in fact, that several teams of researchers and hundreds of hours of surveys devoted to searching its presumed range in northern Alaska yielded only a single data point: one bird. Genetically, gray-headed chickadees are closely related to black-capped chickadees, the commonest of backyard species, which I have also spent half a decade studying. In other ways, they couldn't be more different. Seeing a gray-headed chickadee is special not because its feathers shimmer with iridescence or because it has just arrived from Polynesia but because almost nothing is known about these tiny birds. If I hadn't been paying attention, if I hadn't tuned my ears to the patter of wings and the echo of silence, I would have missed it entirely.

I watch the chickadees as they flit and glean, pulling invisible insects from the needle-clad branches. I take careful note of the shades of gray on the adults' heads and study the contrasting patterns of their feathers, fully aware that I may never see another one of these birds in my lifetime. At the edge of a river that nearly claimed us, I feel the soul-stretching awe that comes with discovery. I feel like a biologist again. Today's rare sighting validates the many late-night computer sessions, the endless hours of packing and planning, every instance of my not feeling smart enough to be a real scientist or strong enough to be a real adventurer. It even makes swimming across the Chandalar River seem like a decent idea. Here, right now, there is only me, Pat, and a family of tiny gray-headed chickadees above us.

Eventually, we leave the birds behind and begin to hike up a steep slope, sweating, our bodies finally warm from within. When we crest a rise, views open broadly into the next valley. The tundra blazes red and yellow beneath our feet. My arms swing more freely with each step, shaking off the morning's scare. Pat's pace is matched perfectly to mine.

For the rest of the afternoon, all the answers I need are in front of me. The sky as big as we are small, our forms dwarfed by mountains and rivers and wide-open spaces. The way Pat and I stop in unison to watch a bear trundle across the valley, each of us reverent and wordless. The scientist in me, having shed the degrees and statistics, once again filled with wonder. The realization that if we weren't doing this, *now,* we would always be missing something.

PART ONE

Zugunruhe

ALASKAN KID

I haven't always loved the outdoors. But as a kid growing up in Alaska, there was no escaping its offerings or denying the fact that the forty-ninth state, with its coarse manners and vast acreage, was home.

Shortly after my parents were married in 1974, they packed an old green Ford utility van and began a three-week journey from Michigan, traveling on what was then an unpaved, remote road that stretched north from Canada to Alaska. Even today, after millions of dollars of road improvements has made it something of an RV thoroughfare in the summertime, bumper stickers still proclaim, "I survived the Alaska Highway." Forty years ago, this boast actually meant something.

My parents weren't dodging the draft, the law, or anything, really. They had only planned on taking a road trip to see somewhere new. But once they arrived in Alaska, they never left. My dad began working for a local engineering firm. My mom accepted a job as a special education teacher. With friends who assumed the role of family, they spent their weekends fishing, hiking, and boating. Though they both grew up in the flat expanse of the American Midwest, they signed up for courses in rock climbing and mountaineering at the local university. Two years later, they spent sixty days on an ascent of twenty-thousand-plus-foot Denali (still dubbed Mount McKinley at the time), after sewing their own clothes and gear, shuttling supplies by dog team, and snowshoeing in from the railroad.

I scaled dozens of peaks in the Chugach Mountains when just a whisper, and then an unwieldy bulge, in my mom's belly. Later, my sister, brother, and I were carted off to campgrounds and remote ski-in cabins by our parents. They signed us up for running races and Nordic ski lessons. In their view, spending time outdoors held the same importance as attending school and playing with friends. A child simply couldn't thrive without it. Now, I couldn't agree more. But at the time, I often resisted.

In elementary school, I buried myself in books. I found an escape in stories, the catharsis of sharing in someone else's triumph or sorrow. I read about the discoveries of Marie Curie and Louis Pasteur. I understood what it would mean to lose a beloved dog from *Where the Red Fern Grows*. I learned that penguins incubate their eggs with their feet and that moose are accomplished swimmers. I found out that the Arctic terns nesting on a lake in our neighborhood had flown from the other side of the planet. I realized I could study almost anything I wanted without ever lifting a foot. Eagerly flipping dog-eared pages and peering through thick glasses, I parked myself in a sunny corner of our living room until my parents forced me to go outdoors.

"You need some fresh air," they'd say. "It's not good for your eyes to read *so* much." I protested these interruptions into the narratives that shaped my view of the world and allowed me to travel thousands of miles from the comfort of the couch.

I would bargain for my reading time—a chapter book for a hike, James Herriot for putting on my snow gear.

For many years, the wilderness didn't speak to me in the same way that books did. I was a chubby, uncoordinated kid with knobby knees and sensitive skin that was easily irritated by heat or cold. When we went camping or hiking, I complained about mosquitoes and steep hills, being wet and uncomfortable. I don't think I minded the conditions so much as the fact that it wasn't *my* idea. Still, I had no choice but to join my family on adventures in Alaska's backcountry. The photos from my childhood show ski lessons and shiny salmon, rustic cabins and moose napping in the front yard. Framed at my dad's office was a picture of my four-year-old brother, perched on the foam seat of an outhouse with the thermometer near his head reading thirty degrees below zero. This was just what we did, no questions asked.

For my parents, who weren't particularly religious, the outdoors offered a version of church that provided clarity without demanding a particular form of allegiance. It would be years before I found the same satisfaction that comes with being part of something larger than myself. I didn't yet see the connections between wild places and wild ideas or understand that my love of animals actually ran much deeper, far beyond their cute and fuzzy nature, and would eventually lead me to become a biologist. Only later would I realize that books were my gateway to learning, but that someday I would need to discover things on my own.

By the time I started planning for college, I had plenty of passion but no idea what I wanted to study. Though I excelled at the calculations and report writing required in my high school math and science classes, I dreaded coming up with hypotheses and designing experiments. There were no instructions, no recipes to follow, and I was always afraid I would make a mistake. I was careful to choose lab partners who had creative ideas; they chose me because I knew how to get the answers right on the tests.

Almost as soon as I arrived at the University of Arizona, which I selected as much for the older boyfriend who went there as for the academic scholarship it offered, I realized the university was wrong for me. The school was huge and freshman classes hosted a hundred students or more. The university's social scene was built around sororities and fraternities, which I had no interest in joining. The pre-med classes I attended were competitive rather than collegial, and I no longer felt like the smart kid, or even smart at all. As I muddled through introductory chemistry and biology classes, shaken by the first B plus I'd ever received, I signed up for a conservation biology class, hoping to find something other than the overcrowded, impersonal courses I'd come to hate.

The first day of class, I met Dr. Bill Calder, a professor in his sixties who had more energy than our entire roster of twenty-five college students. He announced that the voluntary "lab" component of the course would actually be a series of field trips. By the second week of the semester I came to understand what this would mean. At dawn on a Saturday morning, a small group of us piled into a ten-passenger van and headed east to the mountains in search of an elegant trogon—a glamorous, long-tailed bird that rarely strays north of Latin America. Our professor, who quickly informed us that we should drop the formalities and call him "Bill," led the way through poison oak and tangles of juniper bushes, binoculars swinging. Though we never spotted a trogon, and many of us showed up on Monday with itchy, weeping rashes, I was enthralled. We had seen the tracks of a mountain lion, learned the calls of birds I'd never known existed, and found the skeleton of a hawk.

On other trips that semester, we drove south and crossed the border into Mexico, following dusty, rutted roads to reach a riparian corridor threatened by illegal cattle grazing. On some outings, we would get to our destination, climb out of the van, and simply look around for a while before heading back. Other times, we knocked over fence posts, pulled invasive weeds, and dug trenches to return water to the earth where it had been diverted by a culvert. From these trips I learned to look and listen, to get dirty and ask questions.

While we were digging, covered in mud, Bill would wave his arms and holler for us to stop mid-shovel, then he'd point to the sky. "Look, it's a short-eared owl. What do you think she's doing around here right now? No one knows where they spend their winters. Maybe one of you could find out for us."

Bill rose early, hiked fast, read voraciously, and expected the same of us. He'd roust us from our tents just as the sun had begun to color the acacia bushes red, and prod us to *hurry, get up, there's so much to see out here.* And, sure enough, I would crawl out of my sleeping bag to hear the calls of a Gila woodpecker or notice the way the desert grass furled in response to the first hot rays of sunshine. He woke us one night at midnight to check on the cereus cactus we'd identified near our campsite; this drab, scrappy desert plant blooms only a single night each year, exploding in synchrony across the desert with fragrant white blooms. Upon hiking down the hillside, we found the plants still in their withered, unremarkable form. We were weeks too early to witness the cacti bloom, as I'm sure he knew, but he wanted to instill the lesson that we'd never see anything amazing if we didn't look for it. As we returned groggily to our tents, he said cheerfully, "Well, look at Pleiades. Have you ever seen those sisters shining so brightly?"

Formally, Bill studied hummingbirds, but in practice he studied anything and everything about the natural world. He taught us not only about the ecosystem of the Sonoran Desert but about curiosity and wonder. He showed me that, even after fifty years of research, there were always more questions to ask. There was always more work to be done. He also made no secret of his future plans for us. "It's up to you all, you know. These places won't be here if no one loves them. The old farts like me are going to die, and then you'd better be ready to step in." I came often to his office hours, not because I needed help with tests or assignments, but because I loved to hear his tales and imagine myself as passionate about biology as he was. Halfway through the semester, he suggested that I search for summer jobs studying birds in Alaska. "The most exciting projects are literally in your backyard," he told me.

Despite Bill's encouragement, my career as a field biologist had anything but an illustrious start. Though I had applied for more than a dozen summer jobs, I returned to Anchorage the next May with only one prospect. A local ornithologist who, in an odd twist of coincidence, would later become my boss and mentor, allowed a small number of volunteers to help at her bird-banding station. A family friend told me, "She's a great teacher and she might take you on as a volunteer this summer if you show her that you're a hard worker and a fast learner."

All I had to do was be at her field site by sunrise. In south-central Alaska in June, this meant 3:45 a.m. Ignoring my alarm, I overslept and showed up six hours late. The banding had all but finished, the coffee thermos was empty, and the crew of five volunteers looked at me like the pathetic, late-rising teenager I was. Even worse than missing the action, I lied about it. "I must have gotten the directions wrong," I said lamely. "I thought it was on the other side of the marsh."

The field leader only glanced at me, gave a half smile, which I would come to know much later as her *I'm not impressed, but I'm not going to say what I think right now* expression, and said, "Well, maybe you can join us another time." I don't think she remembers me from that day, and I have never seen a reason to remind her.

Desperate for a job, I started working at the local grocery store as a cashier, resigning myself to a summer of studying PLU codes rather than species names. Days later I got lucky. Someone had backed out at the last minute from a field crew, and they needed a substitute volunteer. I had zero qualifications except the fact that I was already physically in Alaska. The crew leader who called to interview me asked, "Can you sleep in a tent and count to one hundred? If so, you're hired."

As part of the required field training, I got a crash course in being a "real" Alaskan. Big-bellied men with decades of hunting and fishing under their belts taught me to operate an outboard engine, drive boat trailers in reverse, stare down bears, shoot guns, and administer first aid in remote settings. Very little of this training had anything to do with birds, and all of it intimidated me. Standing next to the mostly male roster of other attendees, I, a petite nineteen-year-old terrified of embarrassing myself, wondered if I had any business being part of this seemingly macho operation.

Only when I traveled to our field site, tucked inside a narrow glacier-rimmed fjord in the heart of Alaska's Prince William Sound, did I start to appreciate what a gift this season of fieldwork would be. From the deck of a twenty-six-foot boat, I caught my first glimpse of our study species. Black-legged kittiwakes are small, delicate seabirds with clean black wingtips. They breed on rocky islands and spend their winters on the open ocean. Though I'd undoubtedly seen them in Alaska's waters plenty of times before, I'd never bothered to look closely enough to actually identify one. As our boat nudged into the inner bay through a narrow channel, streams of kittiwakes began to pass overhead, traveling neatly in two directions like cars on a divided highway. Their distinctive *eeh-ooh* calls echoed off the steep rock walls of the channel as they flew with apparent purpose. Inside the bay, just a short distance from where I would pitch my tent for the summer, a large piece of rock exposed by the receding glacier hosted the largest congregation of birds I had ever seen.

Suddenly, the entire colony flushed, several thousand kittiwakes taking flight at once. They flew so close to one another that, for a moment, I couldn't see the sky above me. As they came directly overhead, I ducked. When I looked up again, the palette of colors—white wings against blue sky, gray rock against green water—left me gasping for breath. I had never seen motion so synchronized or a setting quite so stunning.

"Peregrine," Rob, one of the longtime researchers, said. Without even seeing the falcon, he knew from the behavior of the birds exactly what had happened. Peregrine falcons are fast and acrobatic flyers. They catch smaller birds on the wing, snatching them from the air in an instant. To evade a falcon, kittiwakes instinctively band together, moving in unison like a sheet flapping in the wind. The more tightly knit each stitch of the sheet, the more difficult it is for a falcon to pick out an individual bird. As Rob spoke, I felt something change inside of me. I wanted more than anything to understand this place, these birds, the way he did.

In the following weeks, I learned that working on a seabird colony demands not only full rain gear—seabird shit is no laughing matter—but a tolerance for high decibel levels. It's a noisy and chaotic scene. Imagine endless rows of miniature apartments, the bustle of fifteen thousand lives crammed into half a city block, each bird working in the frenzy that comes with the short and precious summer. Every minute of every day, kittiwakes were frantically building nests, laying eggs, feeding chicks, defending territories, greeting mates, arriving, departing, sleeping, and, most of all, vocalizing. Soon, I forgot the artificial silence that comes with being indoors. I slept and woke to a cacophony of voices and the sloshing of the tide, accented by the occasional crack of calving ice.

As soon as the first eggs began to hatch, I was assigned the job of tracking the growth of the kittiwake chicks. At the time, it seemed like a compliment to be given primary responsibility for monitoring the chicks, though I realize now that it was nearly impossible to screw up and thus suitable for someone with my lack of qualifications. Twice a week, I boated the short distance to the colony and visited several dozen nests, weighing and measuring each chick as it transitioned from fuzz ball to fledgling. Throughout the season, I studied the intricacies of the kittiwakes' densely occupied neighborhoods and their predictable habits, some pairs swapping parenting duties several times a day, some leaving their nests unattended for hours. I learned that certain birds traveled fifty miles only to feed on the waste products from canneries, the offal that equates to junk food for seabirds, while others found fish in nearby bays. For the first time, I saw the natural world not through textbooks but through my own eyes. I began to understand how ecological questions I'd learned about in school were embedded in the muddy, messy realities of fieldwork, and I loved it.

For each of the following summers, I returned to Alaska to study eagles, sea ducks, shorebirds, and songbirds; during the academic year I helped with local research projects in Arizona. After I graduated, my field seasons stretched longer and longer, and sleeping in a tent, cooking over a camp stove, and working outside in the rain and snow became second nature. The question was no longer what I wanted to do with my life, but what remote site I might visit next or which remarkable species I would study.

NEW YORK
MEETS ALASKA

Pat had a connection to the north that ran almost as deep as mine. The day I met him, he was just a few weeks out of the college dorms, a long-haired roommate to my kid sister. It was early October 2001 and I was visiting my sister in Bellingham, Washington. I walked through the back door of her drafty, high-ceilinged rental house to find a twenty-one-year-old version of Pat at the kitchen table, textbook open, his head buried in a pile of papers. I hobbled in, legs covered in mud, wearing a stained fleece pullover and a winter hat. Several hours earlier I had crossed the finish line of my first marathon, and my body was beginning to rebel. Legs in spasms, I barely glanced up as I said a hurried hello to Pat and headed for the shower.

That night, from the corner of the couch I had taken over for the weekend, I watched Pat flip through slides of a climbing trip he'd just finished in the North Cascade Mountains. From beneath his scratched white helmet, he grinned at the camera, leaving no doubt that the top of a mountain was exactly where he wanted to be. When I asked Pat about the peaks he'd climbed, he was animated and chatty. But on most other topics, he had little to say to me.

The next day, my sister mentioned a cabin Pat had built in the Alaskan woods after he finished high school, when he was just nineteen years old. What, I wondered, would motivate a teenager from New York to find a remote piece of property thousands of miles from home, build a cabin, and spend a solitary winter there? When I probed Pat for details, he shrugged and said it was something he had always wanted to do. He told me how he'd arrived in Fairbanks days after graduating high school with a collection of borrowed goods: an uncle's chain saw, rifle, and boots, and a friend's husky. Just about everyone he met had tried to talk him out of the idea. *But you did it anyway,* I remember thinking.

Later in the evening, we walked in the rain to a downtown brewpub. At first, I flirted with Pat as an experiment, wondering if he reserved his passions only for mountains and woods. I wanted to test the rules of his unfamiliar social conventions. But after a couple of strong Scottish ales, our conversation began to hum with the first twinges of desire. On the dance floor we bobbed to the tunes of a local bluegrass band, grinning at each other in the hot, packed room. Dancing with Pat was a completely new experience. Not in his moves, which he mostly reserved for his upper body, but in the way he looked at me, his face open but unreadable.

Only later did I understand I had witnessed a rare event.

"Pat *never* dances," my sister said.

I left for Anchorage the next morning, full of questions about my little sister's roommate.

Though barely twenty-three myself, I felt much older. I had already finished college and started my career as a field biologist. I was trying to shake off the most recent in a series of boyfriends, all of whom wanted more from a relationship than I did. The last thing on my mind was finding a life partner.

But when I returned home, I could think of little else

besides Pat. What had our brief time together meant? Who was this wonderfully strange person? When I saw a ninety-nine-dollar sale on plane tickets between Anchorage and Seattle several days later, I couldn't resist the temptation to find out if the sparks I'd felt were something worth chasing. A week and a half after Pat and I had first met, I hopped on a flight to Washington, a trip poorly disguised as a sisterly visit.

When I called Pat to tell him I was coming, he said only, "Oh. Wow."

Sheepish about what suddenly seemed like an aggressive pursuit of a person I hardly knew, I began to squirm as soon as I stepped on the plane. But when I showed up, I realized that there would be no awkward, rehearsed script of flattery and flirtation. Pat didn't try to court me with the usual dinner dates or gushing compliments. Instead, we spent much of the weekend lying on our backs by the ocean or running along dirt trails that wound through forests of cedar and Douglas fir.

One afternoon, he reminded me that we had actually met before, when I came to see my sister in her dorm room half a year earlier. I remembered another dark-haired Patrick who caught my eye at the time, but not Pat. Not the shy freshman who would someday become my husband. Not the person who gently grabbed my arm while we were running on the trails around campus and pointed at a rock that he had hoisted into a notch in a cedar tree as an installation art project. The same one who, a decade later, would stop me again at the exact same spot, this time showing me how the striated bark had grown around the granite's gray contours just as he had envisioned.

Before I left Bellingham the second time, Pat showed me a photo of his cabin. The perfect log structure might have been mistaken for the homestead of a settler from another era, each joint carefully notched and chinked with moss. There was a hand-hewn door, an elegant wooden latch, and a roof of sod. Behind its construction lay more care and passion than seemed possible from a teenager.

As Pat and I began to explore the possibility of something other than a weekend fling, we spent more hours on the phone than I had since middle school. At first, I found myself filling uneasy silences with chatter, rescuing Pat from what I thought was a lack of anything to say. It took months before I realized that I was constantly cutting him off. He was slow and thoughtful in his responses, and my impatience meant I was missing much of what he might have told me.

The more I learned about Pat, the more obvious it became that we were, by many accounts, opposites. I was most comfortable in academia. Pat felt at home in the woods. I was a bookworm. Pat could barely spell the most common English words. The first e-mail I received from him might have been written in a different language. It took me fifteen minutes to decipher the two-paragraph message, guessing at the phonetics that, in his dyslexia, he had tried unsuccessfully to piece together. When I told Pat I secretly enjoyed taking standardized tests, he admitted, "When I get those things, I just hope that I'm having a lucky day."

In other arenas, he had a confidence I found completely foreign. I dreamed of climbing mountains someday. Pat simply climbed them. When I asked him how he had learned to scale frozen waterfalls or build anchors on steep rock faces, he looked puzzled, as though the answer hardly needed explaining. "I borrowed some gear and tried it." When he talked about the year he spent building and living in his cabin, I listened quietly. I had little to add. I had never cut down a tree, much less thought about how I would assemble a cabin. Pat didn't boast about what he had done, but his excitement filtered through in his stories of mushing a dog team, being followed by wolves, and dodging wildfires.

I quickly realized that Pat wasn't an ordinary New Yorker. Or ordinary by any measure. As a boy growing up in the Adirondack Mountains, he was the youngest person to climb all forty-six peaks during a single winter. In Alaska this might be equivalent to summiting the foothills of the "real" mountains. Still, for a sixteen-year-old kid, it took dedication, endurance, and a lot of backcountry sensibility. After exploring his local woods, Pat decided that he would head north when he finished high school. He explained that for as long as he could remember, he had planned to build a cabin in Alaska. He'd barely been west of Colorado, and most of his knowledge about Alaska stemmed from books and movies, but it seemed like his kind of place. His mother would later tell me—delighting in the fact he had finally found his way back to Alaska, and with a girl, no less—that even as a young child, he craved the outdoors.

"You borned me in the wrong place and the wrong time," a five-year-old Pat cried to her when he read stories of homesteaders and adventurers from within the confines of his own suburban house, which he shared with four sisters and a brother. They lived in a small town but one recognized more for its well-heeled horse-racing roots than its bucolic nature. His father introduced him to carpentry at their family's cabin, in a part of the country that had long ago lost its pioneers and large predators. Though his siblings were athletic, his family only occasionally hiked and camped together. Still, Pat was sure of what he wanted for himself. He would find his way north someday, to the last great wilderness.

During his final two years of high school, Pat daydreamed about spruce trees, sketched cabin plans, collected necessary supplies, and squirreled away money from his lawn-mowing business. He planned to spend a year in Alaska and then figure out later what to do with the rest of his life. Averse to filling out paperwork and a bit behind the curve on the growing Internet boom, he didn't waste his time with the details of *where* he would build a cabin or whether such a thing might be allowed. This seemed only a minor stumbling block to be dealt with later. When he got to Alaska, he didn't purchase land or stake a claim. He simply found a place with a good stand of trees, received assurance from a local family that no one was likely to bother him there, and started building. Ownership wasn't the point. Adventure was.

He was unlike anyone I had ever met. I was by turns smitten and utterly confused.

By the second month of our long-distance relationship— Pat sharing a house with my sister and other college roommates in Bellingham, me working in Anchorage—I decided I had to see Pat's cabin for myself. I felt like I wouldn't really know him until I did. So, on one of our first "dates," with plane tickets still absurdly cheap, he flew up to Anchorage for a week-long Thanksgiving break and we made arrangements to visit his cabin.

Before driving north, I wanted Pat to meet my friends. But with each introduction, in local coffee shops and on the ski trails, I felt more awkward. The conversations all followed a similar trajectory:

"Hey, nice to meet you."

"Yeah," Pat would reply.

"Are you around for a while?"

"No, just a few days."

"So, you're visiting from Bellingham?"

"Yep."

I squirmed while they made a few more attempts at starting a conversation, usually followed by silence on Pat's end. And more silence.

"Why didn't you talk?" I asked Pat later.

"I don't know, I didn't have anything to say, I guess."

I knew he had amazing stories to tell—I'd heard them. He was witty and funny when we were alone. But with strangers, he went mum. Pat told me much later that he had been intimidated by my seemingly adult lifestyle. I had a car, a job, *and* an apartment. Never mind the fact that the car's interior had been thoroughly mutilated by the previous owner's dog or that it required rolling down the driver's-side window to open the handle from the outside. Or that my job as a biology field technician, chasing birds all over the state for very little pay, was seasonal and temporary. And the apartment? Owned by my parents, who lived in the main house downstairs. I couldn't believe that someone who had built a cabin by hand and spent an entire winter alone would be impressed by these small signs of independence. But for Pat, who had moved out of a college dorm room only six months earlier, I seemed much older than the extra year and a half I had on him. His disinterest in small talk and normal social convention was more puzzling than irritating, though I realized then that there would be no tidy categories into which this enigmatic version of a boyfriend would fit.

Normally, to reach his cabin, Pat would leave from a sleepy community forty miles outside of Fairbanks and follow the Salcha River along its sixty miles of meanders to a tiny island dotted with spruce trees. When he returned to town for supplies, he would stop at the small roadside café, buy a plate of eggs and bacon, and call his parents from an unheated outdoor phone booth. He slunk in and out, dodging questions from the woman who poured his coffee, the curious eyes of the store clerk. "Just staying up the river a ways," he'd say. "An old miner's cabin." No one would see him for weeks or months at a time. Then, out of the woods, he'd appear again. He never said much. No one knew his name. So they called him simply the Salcha Kid.

To travel to his cabin this time, we borrowed an old snowmobile, the two of us piled on the smoky, belching contraption. Behind us, we towed our gear and food in a yellow fiberglass sled that my parents had hauled up the flanks of North America's highest mountain twenty-five years earlier. When the descent became too steep to continue, we found a small flat bench and left the machine there, stashing the keys in the top pouch of my backpack. Pat harnessed the sled to his waist, tugging hard over the uneven ground. I smiled when I pictured my mom and dad, not much older than we were then, with their matching sleds.

We trudged downhill on foot into a patch of black spruce that looked indistinguishable from every other stand of trees in the area. It was only three in the afternoon, but dusk swept swiftly through the forest. Snow filled my boots with each step. I swung my arms in wide arcs to warm my fingers and tried to push away the doubt that began to surface. *Have I been naive to put so much trust in someone I hardly know?* Another three hours passed before we emerged from the black forest to the startling expanse of frozen river below. Pat stood at the bank for a long time, staring at the other side and swiveling his head back and forth—upriver, downriver, upriver, downriver. The snow sparkled under a slender yellow moon. Pat turned to squeeze me and plant a kiss on my frosty cheek.

"This way." He pointed downriver.

Only later did Pat admit that his amazing navigation was, in part, amazing luck. The route we had taken was different from the one he knew so well, but we emerged at a distinct bend in the river just a quarter mile from the cabin. I would soon learn that this good fortune seemed to follow him wherever he went. He touted an optimism that was hard to resist. Sure we could climb that rock face. *What's stopping us? It doesn't matter that we don't have a map. We'll find it.* And, incredibly, we could and did. Most of the time, anyway.

The perfect little log structure was nestled into a tall stand of spruce. It didn't seem possible that a kid from New York would be able to show up with only a few hand-me-down tools and no knowledge of the area and build something like this. It wasn't just a home; it was a piece of art. A local man who had pointed him toward the remote squatter's property where his cabin now stands later told him, "Well, if I knew you were going to build a cabin like *that* I would've had you do it on my land."

When we stepped through the door, the life Pat had crafted and abruptly left only sixteen months earlier sat waiting for him. A hand-carved chair, a double bunk, a small table, and a bookshelf. A stack of kindling lay adjacent to the woodstove.

"Here it is," Pat announced shyly. We dropped our packs by the door, the light from our headlamps dancing across the rough-sawn floor, along the smooth log walls, past the pots arranged neatly beneath the countertop. Other than a few mice and squirrels, no one had been inside since Pat closed the door and boarded up the windows. We lit a fire and soon the frost melted from the log walls of a space that measured no larger than that of a child's bedroom. We devoured two boxes of macaroni and cheese as Pat described his year alone.

The cabin, and almost everything in it, came from local materials. The frames of the bunk beds in the corner were made from spruce saplings. A web of rope strung between hand-drilled holes in the frame formed the two mattresses, each narrower than a normal child's twin.

"Do you want the upper or the lower?" I joked. Pat looked at me for a moment before seeing my smile. I pulled out our sleeping bags and zipped them together, tossing them onto the upper bunk. We stripped down and climbed in. That night, I fell into a secret world shaped by the same two hands that now traced the contours of my body. The world had been distilled to the river, the ice, the trees, this perfect little cabin, and us. As I lay naked next to Pat, our bodies pressed tightly together in the tiny bed, I whispered to him, "I love this place."

Despite sharing what felt like the most intimate of spaces in the most intimate of ways, Pat still largely remained a mystery. I had never spent a week alone, much less an entire winter, and his motivations baffled me. Was he a hermit at heart? I wondered what really lay behind his piercing blue eyes.

In the morning, a gray jay appeared outside the cabin. Winter in the Alaskan interior is a quiet season, with frigid temperatures and little to eat, making any visitors—human or animal—notable. Known colloquially as "camp robbers" for their tendency to capitalize on food left unattended, gray jays belong to the family of corvids. Corvids are among the smartest of birds, recognized for remarkable feats of memory and problem-solving that rival those of dogs and even some primates. Jays can recall the precise location of cached food months later; closely related American crows can recognize individual humans by face. In the case of a well-studied crow population in Seattle, researchers have taken to wearing masks to avoid being recognized by the clever birds. It's no wonder that gray jays need their wits to survive the cold temperatures of the places they call home. On a per-pound basis, a gray jay has a caloric demand eighteen times higher than my own. Using sticky saliva to hide food under bark, in clumps of lichen, and in crevices, the birds handily retrieve their caches throughout the winter and spring.

As we cooked pancakes on the woodstove, the jay waited patiently on a branch by the door as though it knew it was only a matter of time until we emerged with its breakfast. Pat walked to the outhouse and the bird followed, squawking at him. When I went outside, it ignored me entirely. Even after I tossed a burned pancake its way, the jay continued to stay close to Pat, apparently unwilling to swap loyalties so easily. It seemed that it knew a friend when it saw one.

Pat told me that a pair of gray jays had visited regularly during his winter at the cabin. Perhaps this was one of those birds, or their offspring. Even if driven largely by the memory of free food, the nod of a jay seemed like a high compliment indeed, and I took the bird's recommendation to heart.

Over the next two days, I saw how Pat had come to regard this little patch of forest like home. He loved its rawness as much as he loved its subtle beauty. He loved the land not because he had conquered it, but because it was a place that refused to be tamed. I had found someone who appreciated Alaska as much as I did, though as an artist rather than a biologist. He also happened to be the most competent woodsman I had ever met.

In the city, Pat seemed shy and awkward, but outdoors he was mature beyond his years. I could see already that he would challenge me in ways I hadn't experienced before. His vision carried him much farther than remote field camps and wallowing through mud in hip waders with a crew of other aspiring biologists. He wanted to find places where he would be the only human for miles. I was also drawn to wilderness, but much less boldly. Where I dipped a toe, Pat plunged. He was willing to risk everything for a dream. Would I do the same?

It wasn't long before I had my first opportunity to find out.

WHAT CAME BEFORE

I was chasing sea ducks on Unimak Island in the eastern Aleutians when the idea for the first grand adventure with Pat began to take shape. These far-flung volcanic islands extend so far west they eventually cross the International Date Line; on the map they look like a misshapen green necklace against a blue dress, a string of jade splitting the Bering Sea from the Pacific Ocean. Summer offers a haven for nesting birds of every variety, including some that breed nowhere else in the world. Winter brings sea ducks and weather of the sort that has made the Bering Sea crab fishery both deadly and famous, with ice-encrusted boat decks and dramatic rescues that are now the subjects of a popular reality TV show.

It was early February and tiny icicles dripped from the vertically hung mist net that we used to catch ducks in flight. There were four of us working that day, each clad in neoprene chest waders, float coats—insulated jackets that served the dual purpose of keeping us warm and buoyant—and as many clothes as we could fit beneath our outer layers. The clouds shed a mix of sleet and snow and the wind blew hard, with no trace of the tropical warmth of the South Pacific, where the weather system had originated. Two members of the crew drove a small inflatable boat, attempting to herd birds toward the net, which was suspended above the water's surface with floats and aluminum poles. I waited on shore, shivering, poised to wade into the water when needed. Half a dozen harlequin ducks swam just out of reach of the boat, diving underwater to feed on mussels before popping up again far from our net. They glanced at me occasionally, more curious than afraid.

Almost every research project on which I've worked has required catching birds, a task that often leaves me wondering whether it's we or they who are truly capable of "higher thinking." There are as many different capture tactics as there are bird species. Some birds respond to bait, attracted by the promise of free food. Crows and gulls are crazy for Cheetos and hot dog buns. Chickadees favor peanut butter. Mallards come to corn. A few species are so dedicated to sitting on their nests that they can be caught with a long-handled dip net and a casual amble, as long as you avoid eye contact. Others are drawn to decoys and will come to check out the latest intruder in the neighborhood. Warblers will mob a fake falcon. A hawk will land on a captive pigeon. Ducks like company. We're counting on these social habits today, although so far our efforts have been fruitless.

Often, just when I'm sure we've found the perfect strategy, the birds do something entirely unexpected, and leave their would-be human captors humbled and shamed. On one occasion, I crouched in the bushes for an hour, my neck cramped and aching, and watched a crow walk circles around my trap only to realize later that several of its friends had found their way into the bag of bait and happily carted away the spoils to share with the rest of the flock. The teasing bird cawed jauntily at me as it flew away to join its coconspirators. I've seen birds fly over, under, and through holes in my net. I've fallen in the ocean, in a river, in the mud, in a pile of trash, and on the ice in pursuit of birds that got away. Being outsmarted by a bird had become a routine part of my job.

Today was proving to be no exception, as my numb toes could attest.

My eyes were on the ducks, but my mind was on the conversation I'd had with Pat that morning. I'd called him from a grocery-store pay phone before we'd started work, stealing a few quiet minutes away from the rest of the crew. We exchanged stories—Pat about the peak he'd climbed the previous weekend and the tests he dreaded taking; me about the ice-covered rocks and wily ducks. Then, as it often did, our conversation drifted to future plans.

After our time at his cabin, Pat and I had begun to scheme about a summer trip together. We'd been separated for most of the winter; while Pat attended college in Bellingham, I continued to work at field sites or in the Anchorage office, and our conversations were patched together by e-mail and phone. His written messages continued to be so short and punctuated by misspellings that I usually gathered the gist of his subject but little more. That morning, Pat could barely contain his enthusiasm as he told me about his latest idea. He wanted to make a canoe from the materials of a remote northern forest and paddle down a river.

"We don't have to carry a boat. We'll just take the tools we need to build one. Everything else will come from the site. It's so simple, it's perfect!"

I had little interest in the building part, but I loved the prospect of carving out our own little piece of wilderness for the summer. The idea of having time simply to observe birds without the stress of trying to catch them was appealing, as was that of spending two months alone with Pat. After taking orders from crew leaders all winter, the thought of making decisions on my own terms didn't sound half bad, either. As a biology technician, I had the privilege of doing the jobs that no one else wanted. Between moments of excitement when I managed to catch a duck and felt its heartbeat against my palms there were dozens of days spent waiting in the cold, weeks of errands and preparation as I organized field equipment and purchased food.

Sitting on the beach in my rain gear, I thought about what such a trip with Pat would mean. Leaving in the middle of June would require me to give up a season of fieldwork. I'd spent the past four summers searching for nests, tallying the days until the eggs hatched, and watching the young grow and fledge. All of this happened under the surreal cloak of the midnight sun, and often in the company of a tight-knit group of co-workers. It was my favorite time of year. Still, I knew that the work would be waiting for me next year, but Pat might not. As I watched our empty net blow in the wind and counted the minutes until I could change into dry socks, I decided to take a chance.

We eventually committed to Pat's canoe-building idea, more or less in its original form. We selected the Wind River in northern Canada for its remoteness and the fact that we could paddle north for several hundred miles and still reach a gravel road at the end of our journey. To pull off this crazy endeavor, we needed three things: birch trees, tools, and the food to sustain ourselves for two months. Pat collected the necessary tools. I found vegetation maps of northern Canada that promised birch trees at the headwaters of the river and began to prepare the food for our trip.

By the time we left my parents' house in Anchorage in the summer of 2002, we had packed dozens of pounds of grains and several gallons of Crisco vegetable shortening, which boasted the highest caloric value of anything I could find on the grocery store shelves. We did all we could to shave weight off of our expanding loads. We took only the blades of the tools we would need, planning to fashion the handles later from local wood. Pat designed makeshift lifejackets by stitching nylon fabric that we would fill with our sacrificed sleeping pads for flotation. Just before we drove away, we promised my parents we'd be careful and asked them to do the impossible—not worry.

After eight hundred miles of battered roads that coated the interior of my rusty Subaru with a thick layer of dust, we arrived at Mayo, Yukon. Almost as soon as we began hiking, we realized we had gotten many things wrong. Our packs were far heavier than we'd imagined they'd be. Our food rations were frighteningly meager. The old winter road marked on the map was little more than a rutted bog in summer. After hobbling the first six miles, we decided we would have to split up our loads and double carry, covering each mile three times—forward, back, and then forward with a second load.

It took us nearly three weeks of hiking and shuttling supplies to reach the headwaters of the Wind River. Once we got there, it became obvious that the vegetation maps I'd found had never been validated. There were no birch trees anywhere. Our options were to turn back while we still could or to figure out another way to make a boat. One evening, buried in the boatbuilding book we carried with us, I found a brief reference to fashioning a canoe from spruce bark. I read the details aloud, Pat's eyes growing wider with each word. There were spruce trees everywhere. "This will work," he told me. "I know it will."

We set up camp on the riverbank and began to cut trees, collect bark, split wood into ribs, and dig roots for lashing.

Each day we worked until we could barely stand. Our food supplies were far too lean to make up for all of the calories we burned, and we began to feel the effects of slowly starving. By the time we realized we were in trouble, we had no choice but to continue with building. We didn't have the reserves to return the way we'd come.

Desperate for food, Pat shot a harlequin duck, one of the species I'd been studying just a few months earlier. Instead of the idyllic wilderness experience I'd envisioned, I found myself plucking a duck out of season and feeling guiltier with each feather. The only solace came from the fact that the duck was a lone male, and not a breeding female that might have left a brood of ducklings behind. Though he had done the killing, Pat was reverent with the duck, handling its flesh with care. He seemed to understand why I needed to hold the small body in my hands before we dismembered it. Even at our hungriest—pulses pounding, stomachs aching, blackness falling over our eyes when we stood too quickly—we were gentle with each other, focused on the shared goal of making a canoe that had become our only way out. From the absurdity of our situation arose a closeness that I'd never felt with another person before.

Finally, after several harried weeks of bending ribs, steaming roots, and lashing spruce bark to the gunwales, we began our journey down the river in a craft we dubbed *Sprucey*. We must have looked like ghosts from another era, emaciated, dressed in filthy, oversize clothes, paddling a boat that resembled an artifact more than a canoe. But by the end of the first day, we zoomed downriver with an optimism I hadn't felt in weeks—still achingly hungry but elated by the fact that the canoe actually floated. When we dropped our fishing lines into the water and the grayling started to bite, we knew that our hunger had ended. As we paddled, high on our own success and the plentiful fish that filled our bellies, it was hard to imagine being anything other than happy and in love. Forever.

Our days passed like a dream, punctuated by the newness of each moment. We traveled by the rhythm of the river, floating through the bright Arctic night, later napping on the riverbank in the sun. We searched for sheep and bears on the golden hillsides as we navigated gravel-lined channels. One afternoon, huddled in a cave in a rainstorm, we lit a fire and cooked fish over heated rocks. The next day, we paddled through the orange-walled canyon of the Peel River, the biggest water we'd face in our hand-built canoe. We emerged from the wave train and a pair of red-tailed hawks circled overhead. As they shrieked above us, I knew I would never forget the elation of riding the rapids in a canoe made from the land, invincible and humbled at once. Though we didn't exchange a single word, I could tell that Pat felt the same.

Suddenly, anything seemed possible. As the miles carried us closer to our destination, we considered all the places we might go someday. The rivers and mountains we might explore. The dream of doing it together. We sketched out a plan to cross a large swath of the continent's northern reaches under our own power. The details were fuzzy. East to west? South to north? In summer or winter? Each version was more outrageous than the last. We made a pact that we would do a trip of this scale someday. We even gave it a name. *Trans North America.*

After we returned to Anchorage, my family saw our gaunt frames and began a campaign to help us pack on the pounds. For several days, we gorged on fresh fruit and greasy pizza, took two showers a day just because we could, and marveled at the crisp whiteness of the walls, the comforts of the great indoors. We could think of little besides the lifetime of adventures we had imagined together. In our enthusiasm, we created a *Trans North America* folder and stuffed it full of my scribbled notes and Pat's sketches.

But it didn't take long for the novelty of the city to wear off and the excitement about our trip to fade. Anchorage's urban sounds—the train blowing its horn as it passed, the buzz of traffic, music from a neighbor's stereo—began to keep us up at night. We missed the solitude and silence of the river, the sure purpose of each day. As the impending separation of the fall loomed, we began to get short with each other, the uncertainty of our future as a couple soiling the time we had remaining.

One morning, two weeks after we returned to Anchorage and several days before Pat was scheduled to leave for Bellingham, I opened my eyes, startled awake by a dog barking nearby. I lay still in the luxuriously queen-size bed and stared up at a ceiling that stretched impossibly high above me, feeling lonelier than I had in months. All summer, Pat and I had woken up each day side by side, sometimes curled tightly against each other, sometimes linked only by a draped arm or leg, but always touching. That morning, there was a noticeable gap between us. I was pressed against one edge of the bed, Pat against the other. It wasn't intentional, but somehow our bodies knew the words neither of us wanted to articulate.

Do I really love you?

What does a silly trip in the wilderness mean, anyway?

We hadn't argued much since we'd been back, but we had mostly stopped talking about future plans. The private, time-arrested world that had been ours didn't mesh easily with the realities of everyday life. Surrounded by the bustle of the city, our canoe trip seemed increasingly distant, an experience that belonged to different people in a different time.

I cried when Pat left, mourning what already seemed lost. Like the canoe we built, the bond we had forged felt antiquated and trivial. He finally called me on a pay phone more than a week later and two thousand miles away, and I barely recognized his voice. That semester, Pat had decided to live in a canvas tent in an alley near campus, and the only phone he could use was at a Chevron station half a mile away. I would listen to his messages and call the pay phone number only to hear it ring unanswered in the parking lot, cursing the fact that, for once, he couldn't just be like everyone else and live in a regular house with a regular phone. I felt our love quietly slipping away.

When I flew down to Bellingham several months later, I didn't know what I might find. There was Pat, at home in a tent in the city. His alley neighbors included dogs, trash barrels, and a pigeon coop that belonged to the owner of the adjacent house. He told me about the pigeons as we walked past their mesh-enclosed structure.

"They're homing pigeons," he explained. "No map or compass and they can find their way back to this coop from the other side of the country. That's a whole lot better than either of us could do." Pat was not a birder or a biologist, but he knew enough to appreciate what wonders even these most humble of birds could offer. How could I not love a man so modest that he was willing to share his living room with a flock of pigeons?

The inside of Pat's tent felt much like his cabin on the Salcha River. He had built a bed frame, a desk, and a countertop with a sink that was sourced from a neighbor's hose. There were several photos of us tacked to a small bulletin board. Lying cramped against the damp fabric on his narrow single mattress, I marveled at the complete rejection of convention that had impressed me when I first met my sister's ponytailed young roommate. Though the alley's nighttime noises kept me awake and we struggled to keep from pushing each other out of bed, I found myself settling easily into the same private world we had cultivated on the river. And I suddenly knew that the relationship we'd discovered the previous summer was real, more real in fact than anything else I'd ever felt.

The next semester, I moved to Bellingham and Pat and I began our first foray into sharing a life somewhere other than on the banks of a remote northern river. I enrolled in a graduate program in writing while Pat finished his fine arts degree. We found a deal on a twenty-seven-foot sailboat and became live-aboards in the same marina from which we would launch our rowboats eight years later on the *Trans North America* trip. Each night, we nestled into the V-berth in the boat's bow, the space so small that only one of us could lie on our back at a time. If we needed to change positions, we did so in tandem. As the wind howled and the halyards whistled and we sloshed around in the waves that washed over the breakwater, the annoyance of not sleeping was overshadowed by the exhilarating power of the ocean.

Once again, as we had on the Wind River, we began to imagine a future together.

NOW IS THE
ONLY TIME

After Pat and I finished our degrees in Bellingham, we headed north, with our *Bird Songs of Alaska* CD on repeat. I had a season of bird surveys planned, with Pat lined up as my field assistant. We traveled on foot and by kayak to reach the remote sites, where Pat could stare at glaciated peaks and record data while I looked and listened for birds. It was the sort of fieldwork that suited us both. Later in the summer, after the surveys ended, we drove to Haines, a community of two thousand nestled at the northern end of the Inside Passage. In college I had spent a carefree summer studying bald eagles in this picturesque coastal town. Independent of my recommendation, Pat had pored over maps of the entire northern Pacific coastline and mentally circled the striking interface between mountains and sea. We had come for a kayaking trip, but before we launched, we browsed the window of the real estate office in town.

One piece of property caught our eyes. Described as "remote beach front on Glacier Point," it showed stunning views of a toothy, snowcapped mountain range and described a glacier within walking distance. We wrote down the relevant details on a scrap of paper and grabbed a plat map before leaving, with the casual thought that perhaps we'd take a look if it happened to be on our way. We'd now been together for nearly four years and were ready for something new. More adventure, certainly. But also a home of sorts, a piece of earth to call our own.

When we arrived at Glacier Point, a large glacial outwash plain that fanned into Lynn Canal, a deep, glacier-carved fjord, we stopped for the evening. As we relaxed around a blazing driftwood fire and watched the moon rise over the mountains, we had no idea we were sprawled on a stretch of beach that would tie us to the canal in every way imaginable. The next morning, we crossed a grassy meadow dotted with wild strawberries, ducked through a thicket of young spruce and alder, and popped out in a forest that might have housed a village of gnomes. Light filtered through the old growth spruce canopy onto a mossy floor painted a surreal shade of electric green.

"This is the sort of place I dreamed about as a kid," Pat said as he sized up the trees, noting their height and straightness, the idea of a cabin already taking hold. I, too, loved the prospect of having a plot of wilderness to truly *know*. Not just a place to drop in as a visitor for a week or a season, as I did with most field jobs, but somewhere to plant roots. The decision to paddle back to town instead of continuing on our kayaking trip was an easy one—all of our attention had been diverted to figuring out how to make this magical place ours.

When the sale finally closed seven months later, we began the transition from dreaming to planning, and decided to use trees on the property to build a log cabin. I turned down several biology jobs so I would have my summer free to start construction. Tree by tree, we cut and hauled and stacked the walls that would become our home. Using a hand-built crane, a pencil, and a chain saw, we hoisted and scribed each log so that it nested perfectly into the next. Whenever I could drag Pat away from the building site, we explored the adjacent forests and beaches. We walked past a tidal lagoon, where birds and bears gathered, or hiked the three miles to the Davidson Glacier, nestled against a steep hillside dotted with mountain goats. Even as we worked, wildness often found its way to us. One afternoon, a wolf surveyed our progress from the shadowed corner of our building site. A pair of goshawks fed on deer mice that scurried across the clearing. Bears grazed the field of wild strawberries on our beach.

But as much as we loved Glacier Point, we soon discovered that practicality and passion were at odds. The same remoteness that had first attracted us to the property also made it impossible to live there year-round. We needed an income, even if only a modest one, and our cabin was many miles from town and employment opportunities. Thanks to the portability of laptop computers, I had managed to work part-time during the summers, but this was just a temporary arrangement. I was also anxious to move beyond seasonal jobs as a technician and take on my own research as a graduate student. We compromised by splitting our time between Glacier Point and Anchorage, drawn to the city by work, academic opportunities, and the company of friends and family, but always eager to return to our cabin each spring.

During one of these winters in Anchorage, I became obsessed with an ecological mystery so intriguing it might have come from the plot of a futuristic movie. I was writing a report about the previous year's bird surveys when I first learned about the problem. A large cluster of resident birds in Alaska had turned up with grotesque beak deformities and no one had any idea why. Coincidentally, my boss at the time was not only the same person whose banding station I had failed to "find" eight years earlier during my first attempt at biology fieldwork, but also the lead researcher on the beak deformity project. She was always in need of extra help catching black-capped chickadees, the species most commonly affected, and I leaped at the opportunity to join her.

The instant I saw the first afflicted bird, which sported an appendage that looked like a stick growing from its beak, I knew I had to find a way to be a part of this study. When I asked, my boss told me that there were no funds available, the research had met many dead ends, and she didn't think it would be a good graduate project. This only made me more determined. I had long been drawn to the work of Rachel Carson and other early conservationists who helped the world see how wild animals, and birds in particular, were more than just our proverbial canaries in the coal mine. They were living proof of the state of our environment. I wanted desperately to find out what caused the deformed beaks and what the birds might teach us about ourselves. If a cluster of beak deformities—which are often linked to contaminants in the environment—suddenly appeared in birds that spent their entire year in Alaska, what did that mean for the rest of us?

The next summer, between helping Pat to hoist logs and set windows, I sat on the beach and wrote the proposal that would form the basis of my graduate work. As I chopped kindling or showered under a spruce tree, I thought of chickadees with twisted, overgrown beaks. That fall, I applied for research grants, arguing that problems among these small, familiar birds might foretell something much larger about our environment. Six months later, I was shocked when I received an e-mail stating that I had received the requested funding. Soon after, I began my studies at the University of Alaska, full of enthusiasm that couldn't possibly last.

As I outlined my research project, I scoured the scientific literature for clues and learned everything there was to know about the inside of a bird's beak, which wasn't much. The hard outer surface of the beak is made of keratin, the same material as human fingernails. Beyond this, almost no information existed about the structure of these tissues or the nuances of their growth. Thus, when a beak grew dramatically out of control, as was the case for the chickadees, any hypothesis consisted of little more than a wild guess. My advisors steered me toward a project that would yield clear results, and I was told I needed to focus not on the big-picture questions of contaminants in our environment but on the minutiae of keratin growth. Soon, I found myself plopped in front of a microscope and observing chickadees in cages.

I am not a laboratory person by nature. I can be precise when necessary and attentive to the details that matter when conducting experiments, but stick me in a windowless room for too long, and I begin to take on the same empty expression I saw reflected back at me from the chickadees' stainless steel cages in the university basement. Each day, I dreaded pulling my lab coat from its hook to begin another session of staining glass slides or peering through a microscope for answers that never came. I had taken two dozen birds from the wild into captivity in hopes of discovering the cause of the deformities, but so far I had failed. Perhaps worst of all was the fact that I was gradually coming to despise what had started as an act of love. I cared about wildlife, and the well-being of our environment, but my research left me feeling distanced from the natural world, immune to its wonders. In my mind's eye, the chickadees had transformed from tiny marvels who could survive nighttime temperatures of fifty below zero into lab rats who refused to yield answers.

Pat, though more grounded by nature than me, less prone to questioning every minute decision, was also becoming restless. While I was pursuing my Ph.D., he took on the equivalent of an informal graduate program. With a background in art and an aptitude for building, it wasn't surprising that he had found his way into residential construction. But he didn't merely learn to swing a hammer and work for an hourly wage. In just two years, he transitioned from apprentice to owner of a design-and-build company. Despite his aversion to computers, he taught himself how to create architectural drawings and design homes from scratch, the products of thousands of clicks of the mouse and tiny balsa-wood models. He managed employees and subcontractors, budgets and permits. Perhaps most impressive of all, he convinced clients to trust him, a baby-faced twentysomething with no official architecture credentials and only a limited building résumé.

With the same intensity that had swayed the opinion of the initial skeptics of his Salcha River cabin, he promised he could build a house better, cheaper, faster, and more beautiful than anyone else. And he did, often under ridiculous circumstances. The winter we lived in Fairbanks, he poured a foundation as the snow flew, and installed windows at twenty degrees below zero. One of the temporary laborers he hired later told him that he had only answered his ad on Craigslist to meet the lunatic who would attempt such a thing. But these efforts cost him. His focus had to be complete, leaving little time or energy for the adventures that we both missed.

As I entered the fourth year of my Ph.D. program, I increasingly questioned where my studies were taking me. I had learned by then that passion and research don't always converge. I dug through dusty old journals to find articles on horse hooves and sheep's horns and wondered what I was accomplishing. Arguably, I had made contributions that advanced our knowledge about these sick birds, but I hadn't fixed anything. I began to publish my results, but nothing changed. I worked as hard as I could to find answers, but the ones I discovered didn't seem to make a bit of difference. The birds still had terribly deformed beaks and we still didn't know why.

Perhaps as an antidote to the stasis of laboratory and computer work and the frustrations of not finding the answers I wanted, I ran dozens of miles each week on the trails around campus, sometimes adding an extra hour to my workout just to avoid having to return to the lab. As I combed the scientific literature for clues about keratin growth that I might have missed, I found myself searching by keywords unrelated to my research. "Migration + endurance + Arctic." "Long distance + shorebird." "E7 + Alaska." Several colleagues had recently documented the longest single flight ever recorded for a bird—a female bar-tailed godwit had traveled nonstop more than seven thousand miles from Alaska to New Zealand, reaching her destination in just over eight days. E7, named for the alphanumeric code on her leg flag, became an instant avian celebrity, and I quickly joined her fan club. I pored over the satellite maps provided online of this now-famous bird. I reveled in the collegial enthusiasm of a new discovery. As I stared through the microscope or weighed tiny sample capsules, I wondered why I hadn't studied migration instead of beaks. When spring came again, my eyes drifted constantly from my computer to the sky.

I felt, more than ever before, a kinship with birds in the springtime as they bide their time, waiting for the perfect moment to launch on a journey that will take them across continents or counties, over oceans and forests. For birds, the urge to move can't be contained. Its pull is so intense that a sandpiper's organs atrophy to accommodate the demands of migration. Its siren song lures godwits from New Zealand to Alaska. A caged robin will launch itself northward again and again, hammering against glass walls, even if it has no view of the outdoors. There's a word for this: *Zugunruhe,* a German noun made up of two parts: *zug,* "to move," and *unruhe,* "anxiety" or "restlessness." It means migratory restlessness, and is seen in caged birds prior to the onset of the migratory period. There's no mistaking the signs. Wing fluttering. Sleeplessness. Disruption of normal activities. I had *Zugunruhe* in a big way.

Around the same time, I faced a series of personal losses that seemed to signal a shift to someplace other than the carefree days of my twenties. It was a transition to adulthood I wasn't ready to embrace.

For several months, my mom had traveled back and forth to Portland to provide hospice care to one of her sisters, my favorite aunt, Claudia. The last time I spoke to Claudia, I was standing on the beach near our cabin at Glacier Point. The wind was blowing and the clouds scuttled by, casting shadows on the turquoise water. My aunt had been battling lymphoma for years, but I didn't understand how sick she was until near the end. When my mom left a message to say that my aunt's health was worsening and that I should call soon, I realized I had no idea how to talk to someone who was dying.

As I scuffed my foot against the rounded beach pebbles, I told her I loved her. When she asked me how I was, I described our cabin project. I said I hoped she could visit us someday. Even as the words formed in my mouth, I knew this wasn't possible. There are so many other things I wish I had said instead.

When I was eleven, I remember staying up all night with you to paint a bedroom yellow. We sat on drop cloths, our hands and clothes covered with paint, and drank diet soda. You made me feel so grown-up.

I have always loved the way you laugh, just like my mom.
I still have the silly drawing of the dog you made for me in
the waiting room while my sister was being delivered.
I'm sorry this has happened to you.
Goodbye.

A few days later, she died.

The next summer, my dad's best friend drowned while they were on a rafting trip on Alaska's Kenai Peninsula. It happened on a river I had boated with my dad as a girl. The same river that my dad and his friend had rafted together more than a dozen times before. It was just bad luck that he was tossed from the boat shortly after he had taken off his too-small, borrowed life jacket. It was just bad luck that he got caught in a whirlpool he couldn't swim out of. This jolly man whose family had shared every Christmas Eve with ours for as long as I could remember was gone. His laugh that filled a room and made everyone around him laugh, *just because,* had vanished with the current.

And then there was my dad's health. He had begun to stumble at odd times, his right hand shook, his gait was slightly stooped, and his normally exuberant smile had started to wane. The signs came on gradually, but they soon became impossible to ignore. He finally went to a doctor for an assessment.

The morning I heard the diagnosis, sun streamed in the upstairs window and bathed the room with the optimism of an endless Alaskan spring day. My mom was wearing shorts and flip-flops, her dark hair pulled back away from her face. Even pushing sixty, with laugh lines etched deeply, from a distance she could easily have been mistaken for a teenager—her quick steps, the way her ponytail bounced against her neck, her tendency to go barefoot before most people in town had shed their jackets. But her body stiffened as she turned to explain what she knew. She looked at me sadly, as a parent might when a child demands the truth about Santa Claus long after she should have stopped believing.

"The neurologist thinks it's early Parkinson's."

"What?" I replied.

"He's still on the young end; his aunt was much older when she was diagnosed. But because his symptoms aren't too bad yet, he decided he didn't want to start any medication right now. They'll just keep watching him." Her eyes were full of sorrow and apology, as though she should be able to protect me from this awful news. I blinked hard but the tears came anyway, wetting my cheek as I turned toward the window. My mom took a step closer and hugged me. I hugged her back awkwardly.

Only later would I lie on the floor alone and sob. When Pat returned home in the evening, I began to shake as I tried to tell him what I had learned, then ran into the bedroom and closed the door, covering my face with a pillow against the bright evening light.

When I heard the diagnosis of Parkinson's disease, I knew only that it was bad, but a quick Internet search left me struggling to breathe. Degenerative. Neurological. No cure.

Suddenly, everything around me felt fragile. Our physical bodies no more solid than air. Our plans and aspirations just a way of deceiving ourselves into believing that the future is infinite. My dad, who had climbed the continent's highest peak and done all the things we're told will keep us fit and healthy, now had a body that had turned on him. I wanted both to curl into a corner and weep and to escape to a place far from reality, while I still could.

One April afternoon in 2011, I opened the file cabinet at our Anchorage apartment in search of an undergraduate transcript and came across the *Trans North America* file Pat and I had created eight years earlier. The tab, labeled in Pat's barely decipherable handwriting, was faded and creased. Inside the folder was a collection of notes that told the story of our imaginary journey. A sketch of an umiak, an open boat with a sail originally designed by Inuit whale hunters, that Pat had thought we might build and live beneath during an Arctic winter. A crude map, showing Alaska and the Yukon, with a pencil line stretching from the Aleutian Islands to the Arctic Coast. A page of notes about food and gear we'd used for the Wind River trip. And the most telling piece of all, a photo of Pat and me in Fort McPherson, where we had finished our canoe trip, standing together next to our spruce canoe. We grinned at the camera like the children we were, with nothing but the future ahead of us, each holding a paddle we had whittled into shape and the other's hand.

Something shifted in me that day, even after I resumed my search for the transcript. Our *Trans North America* dream wasn't just a fantasy we might entertain someday. It was something we had to do. Not eventually. Not tomorrow. Now.

When Pat and I began to sketch a route for our trip, we started with the places we knew: The Wind River, where, only a few months after we'd met, we almost starved but instead fell in love with the lure of adventure. The glaciers near our cabin on Lynn Canal that overlooked the sea. The bay in Washington where we'd lived on a tiny sailboat together. Then we began to fill in the places between. The ones we'd always wanted to see. The ones we'd never considered visiting.

Just as important as where we would go was how we would do it. We wanted to experience the landscape as the birds and caribou did: entirely under the power of our own muscles, without using motors, roads, or established trails. We would leave behind clocks and schedules, jobs and commitments, to follow a route that had never been mapped. As we pored over maps, the details changed but the idea stuck. One foot in front of the other. One paddle stroke after the next. The steady swish-swish-swish of skis. A departure from our ordinary lives for half a year, or more.

Our dream was simple, the scale completely outrageous. We would cross four thousand miles of roadless, trailless terrain through a landscape where glaciers are larger than entire countries, oceans nudge the edges of continents, and rivers flow north to the Arctic. The distance we hoped to travel was equivalent to that which separates New York from Stockholm. Such mileage for a godwit, goose, or hummingbird, even, was a migration that came with each season. But for us, as we sketched out a route that grew from a pencil line on a scrap of paper to a map that covered our entire living room wall, it meant more miles than we had traveled in all of our previous wanderings combined. To do this as a single, continuous trip, we would have to push the limits not only of our own bodies, but of the seasons. We would start in early spring and travel through the brink of winter in the Arctic. It was a journey that promised to test us in every way possible.

Planning a four-thousand-mile trip through the wilderness wasn't a logical response to becoming disillusioned by research or facing my dad's illness. But logic was the last thing I needed. These weren't matters of the mind but of the heart, and I could think of only one way forward. The advice my parents had given me for years now rang true. *Go outside, kid. Fresh air will do you good.* The time had come to take on the grand adventure Pat and I had imagined many years ago.

PART TWO

Inside Passage

PREPARATIONS

We're scheduled to leave in four months, and the to-do list taped to our refrigerator stretches almost to the floor. There isn't a single task that can be checked off with a quick run to the grocery store or local outdoor shop. Instead, the list is peppered with jobs that will each take weeks or months of work. "Build rowboats." "Buy food." "Pack food." "Route info." Earlier in the winter we came to the disappointing realization that expedition-style rowboats were not commercially available and that we would have to build them ourselves. Pat balked, then took this on as a night job. Because we've always liked our independence when traveling, and because having two boats is safer than one, Pat is doing each step in duplicate—for my boat, and for his. But all he's finished so far are the plywood shells. I've planned out two hundred days' worth of meals, but I haven't started shopping yet. Our notes for route information contain more questions than answers. I finally get so tired of staring at the list that I pull it down, type the words into a spreadsheet, and congratulate myself for actually accomplishing *something*.

The list is my domain; in all the years I've known him, Pat has never regarded a to-do list as a worthy endeavor. Perhaps it's how he maintains his optimism, working as hard and as fast as he can, dreaming only of the outcome, not the possibility of failure. Or maybe he simply knows that I'll take care of it. My jobs as a biology technician have given me a lifetime of logistics practice. I've planned dozens of meals for strangers. I've walked out of a bulk grocery store with five hundred Snickers bars. I've boxed and labeled and shuffled gear on pallets until I wondered why I didn't sign up to be a longshoreman rather than a biologist. Thanks to my detailed list-making, I've rarely overlooked anything essential. Never, despite one boss's greatest fear, have I forgotten the field pencils.

Sometimes I resent my de facto role as the planner in our relationship, the one responsible for schedules, finances, and research. The one who sees clearly just how far we are from being ready. But Pat's response when I gripe about this pattern is hard to refute.

"Do you think it would be better if both of us tried to be in charge?" he asks innocently.

The scale of our trip is not only daunting; it's a logistical nightmare. We need gear to row, ski, hike, packraft, and canoe. We figure on being gone for at least six months, maybe more. To feed ourselves for this long, we'll need a thousand pounds of nonperishable food purchased, sorted, and crammed into ziplock bags. With the help of my parents, we will mail most of our supplies to ourselves general delivery through the United States Postal Service and Canada Post. Seventeen resupply packages must reach us at exactly the right time and place. For the first twelve hundred miles in our rowboats, we can stash extra days' worth of food in the large hatches without much added burden. But once we transition to ski and foot, our distances will be limited by how much we can carry on our backs. The fact that there are few towns along the way where we can resupply only adds to the challenge.

Months ago, before the reality of packing and preparing had set in, I had focused on the trip as an escape. As I peered under a microscope and stared at graphs on a computer screen, I daydreamed about the freedom of traveling with only what we needed on our backs. I ignored the fact that this freedom might actually translate into some of the hardest work I would ever do. I overlooked the tedium of hundreds of hours of sorting dried food, making phone calls, and weighing every single item we planned to carry for the next six months. I underestimated what it would take to build and outfit rowboats.

Now that our departure date is quickly approaching, I'm frantically dividing my time between statistical models and a tally of calories and miles. I'm scrambling to finish my dissertation as Pat powers through a house remodel that will help fund the trip. When I work especially late at my office cubicle, I call Pat to hear that he is applying fiberglass to our rowboats or wiring the basement of the house he is building. When we both make it home, exhausted, a second job awaits. There is always another task to cross off the list. The trip has become one more impossible deadline to meet.

Many evenings, I park myself in a room in my parents' basement, which we have taken over for our food packing. On those nights, dinner consists of a buffet of snack foods. In a plastic tote the size of a clothes hamper, I dump dozens of pounds of nuts, chocolate chips, and dried fruit. I secure the lid and shake the tote to shuffle the contents into a semblance of a trail mix. First calculating how many calories are in a pound, I weigh each bag, designating two- or three-day rations. Chocolate bars are stacked ten high around the room. Fifty pounds of pasta and a bucket full of couscous occupy one corner. We have a dozen tins of freeze-dried vegetables that were prepared with the end of the world in mind; purchased from a religious retailer, they comply with the biblical mandate to keep a stockpile of nonperishable food on hand. They also happen to taste good in dehydrated soups and mashed potatoes.

On the bookshelf sits my victory stash: several cases of instant coffee packets and two dozen boxes of tea. Pat and I argued again and again about whether or not we should have hot drinks on our trip. With the added time and fuel needed to heat water, a quick granola bar would be the most efficient breakfast. But this is the one luxury I refused to sacrifice, knowing that the familiar morning ritual will carry me through each day.

One evening, my sister, seven months pregnant, helps me package instant oatmeal in ziplock bags. By the time we're finished—eight hundred packets later—we're covered in a fine dusting of sticky oatmeal powder and nearly delirious. For the next six months, the smell of maple and cinnamon will remind me of my sister, belly bulging, sprawled on the floor pouring oatmeal flakes into plastic bags. I will remember how we laughed together that night. I will remember how fortunate I am to have such support.

But even with help from friends and family, we're not making headway fast enough. Pat and I spend an entire day going through the dreaded list, trying to figure out how to be more efficient. I create a giant timetable of what needs to go where and on which date. Pat tapes dozens of topographic maps to the wall and traces our intended route on each of them. When the maps begin to tilt crookedly, I snap at Pat to be more careful before he calmly informs me that it's not his sloppy taping job, but the curvature of the earth that's responsible. The scale is *that* big.

As the days speed by, we focus our attention on the most critical sticking points—the aspects of the route or pieces of equipment that, if not addressed now, are most likely to cause us to fail. For planning purposes, we've divided our route into nine major chunks. For sanity's sake, I've also mentally organized them according to the birds:

INSIDE PASSAGE: Row north along the Pacific coastline. Start early spring. <u>Pat must finish boats.</u> *Scoters feed on the herring spawn. Hummingbirds cruise north.*

COAST MOUNTAINS: Trade rowboats for skis and packrafts. Cross glaciers into Canada. Paddle to Whitehorse. <u>Check aerial photos. Call pilot.</u> *Thrushes arrive in southeast Alaska. Swans congregate on Marsh Lake.*

YUKON RIVER: Rent a canoe in Whitehorse. Float to Dawson. <u>Look at trip reports.</u> *Warblers sing their hearts out. Chickadees sit on eggs.*

YUKON: Hike Tombstone Mountains. Go east to Wind River. Retrace our route to McPherson by packraft. <u>Need a resupply on Dempster Highway.</u> *Songbirds quiet and secretive on their nests. Raptors feed young.*

MACKENZIE DELTA: Paddle to the coast. Mosquitoes. Slow current. <u>Can't get our rowboats to the Arctic. Use packrafts? Is this crazy?</u> *Duck mania.*

ARCTIC COAST: Hike and packraft along the coast of the Arctic Ocean. Everyone says bears are hungry up there. Pack ice? <u>Look at satellite images.</u> *Colts (sandhill cranes) and cygnets (tundra swans) find their wings.*

BROOKS RANGE: 1,000 mile traverse. Maps hang crooked on the wall—Pat claims he's not going crazy. <u>Need to figure out our route!</u> *Most species begin to head south. Birdsong disappears.*

NOATAK RIVER: Almost there! 425 miles to Kotzebue. Nowhere to resupply. <u>Need to arrange an air drop.</u> *Geese stream past. Swans stage on the coast.*

* * *

As we outline each section, we're careful not to state our itinerary too boldly to others. We tell friends and family that we'll start in Bellingham and make it as far as we can. We've realized by now that the trip we have proposed is on par with what only a handful of elite adventurers have managed to pull off. It's an ambitious goal, one larger than any we've ever tackled. But instead of the résumés of professional explorers, we have the credentials of a burned-out biology graduate student and a carpenter. We have no logistics coordinator. We don't have a single sponsor. Never mind the fact that we've hardly paddled white water. Or that we have virtually no rowing experience. Or that we will cross some of the most remote and bear-dense landscapes on earth.

When I worry over all of the things that could go wrong, claiming failure before we've even begun, Pat reminds me of our history. "We've never done things by a guidebook," he says. "Why would we start now?" I recognize that familiar glimmer in his blue eyes, the one that burns most brightly with plans that others deem crazy. I've learned by now that he sees little need to prove himself. Other people can think what they want.

Still, I wonder if we've taken that reasoning too far. We've thrived on the challenges of the outdoors, but we've also become accustomed to accomplishing what we set out to do. Somehow, we've managed to achieve nearly all of our previous goals, no matter how difficult or absurd they seemed. *Sprucey.* Climbing mountains. My Ph.D. Pat's house projects. What if this time we simply can't do it? It's easy to love the wilderness when the wilderness loves you back. Everything changes when you feel like the land could swallow you whole.

When I come across photographs of the Noatak River in northern Alaska in an old *National Geographic*, Pat and I take a break from our list and sit on the couch to look at the article. We press our shoulders together and flip through the pages.

Grainy photographs show a man and his dog in the summer of 1989, tracing their way across the Brooks Range through many of the same areas we plan to traverse. The landscape is exactly how I had imagined it, endless tundra-covered ridges and valleys thick with caribou. Basking in the low-angle sun, this patch of earth sings from the page. I hear the early-morning chorus of Lapland longspurs and Savannah sparrows, Arctic warblers and gray-cheeked thrushes. I feel the pulse of energy that arrives with summer's short glow.

But I shiver when I see the next page. Several photos show fresh snow on the ground and the man's finger black with frostbite. The seasons moved faster than he did.

"Look, he spent the winter here!" Pat says as he points at the map. "Well, we know it's possible if we get a little behind schedule." I look over and see that he is grinning hugely.

The prospect of spending an accidental winter in the Arctic isn't particularly funny, but I can't help laughing when I see Pat's face. There's some part of each of us that is inexplicably drawn to the idea.

The excitement of the unknown tugs hard at us then, even from the comforts of a saggy red couch in a warm apartment. The man on the Noatak River with the frozen beard is a reminder of why we've committed to this trip. Our pact. The magic we first discovered on the Wind River. The bond we forged by sharing food, body, work; the way we chased our canoe-building dream with such focus that everything extraneous disappeared. The way it felt to see the world as though peering through binoculars for two months, with a field of view so narrow yet so clear.

* * *

Those last days before we leave Anchorage, every hour matters. Realizing too late that we have the wrong schedule for the freight barge headed to Washington, we recruit a neighborhood crew to help us package the boats for their journey south. Pat finished them only days ago and their paint is barely dry when we drop them at the shipping dock ten minutes before the midnight deadline. I plead with the agent to let the boats onboard and then sneak into the warehouse to find a forklift driver who can assure me they won't get crushed in the loading process. Before they have come anywhere near water they're hoisted into a cargo container, with large "FRAGILE" stickers plastered all over their sides.

Later that night, I complete the final paperwork for my Ph.D. graduation. It's nearly 4 a.m. by the time the files are scanned and ready. I press send and close my eyes. There's no celebration, only the worry that I might have made a mistake on one of the forms. Chickadees are far from my thoughts.

Our final gear packing quickly becomes a matter of triage. Things we debated about for weeks are reduced to *yes, no,* or *too late to worry about.*

Some of these decisions are easy. Yes to the satellite phone. Besides the extra safety margin it allows, saving our families any unnecessary worry is worth the weight and expense. No to the gun. It's logistically complicated—until recently, firearms were banned from national parks in the United States; they are still illegal in Canadian parks—and too heavy to justify. Despite warnings about hungry Arctic bears, we've generally been of the opinion that a gun is a poor substitute for caution. Bear spray has proven effective in most cases. Yes to the trekking poles for the hiking portions. Our knees will appreciate the support through hundreds of miles of tussocks and scree. No to a spare paddle for the packrafts. We simply can't afford the extra pounds. Yes to the small solar panel that will allow us to charge our electronics along the way. No to dry bags that are supposed to keep the smell of food sealed in and away from the prying noses of bears. The bags are pricey and difficult to find, and in our brief trials didn't seem to stand up to more than a single use. Yes to the cell phone for checking e-mail as we pass through towns.

Finally, our chaotic piles begin to take shape.

LAUNCH

Five days after shipping the boats, we scurry around our Anchorage apartment in another midnight frenzy, shoving last-minute additions of toothpaste, spare boat parts, and neoprene socks into already bursting luggage. Our morning flight to Seattle is scheduled to leave in less than six hours, and the living room still looks like a yard sale. Queasy with fatigue and hunger, I pick up a bowl of cereal I poured for dinner that has sat untouched for hours. Pat joins me at the table and we stare blankly at each other, shaking our heads at this madness of our own devising. As I survey the mess of clothes and gear, I hear a distinctive crack followed by a low, quiet moan.

Pat scowls as he fishes around in his mouth and then spits out a quarter-inch silver filling that had, only moments before, been attached to his molar. "Fuck. You have got to be fucking kidding me." I take one look at the jagged piece of metal, a bit of tooth still attached to it, and begin to laugh before bursting into tears.

Somehow we drag ourselves onto the plane later that morning, begging coffee refills from the flight attendant each time she passes. When we touch down in Seattle we retrieve our luggage under the fluorescent lights of the baggage claim and meet my brother outside in the slushy street. We have landed on one of the rare days when snow is falling in Seattle, and his heavily loaded sedan struggles through the unplowed roads. Later that morning, Pat goes to a neighborhood dentist, returning an hour later with the report that a repair would require more than a single visit.

"He didn't think it was safe to do such a major job and then send me out on a boat," Pat explains. "So I guess I'll just have to chew on the other side of my mouth for a few months."

In the morning we rent a twenty-four-foot U-Haul, collect our rowboats from the freight dock, and wind our way out of Seattle's narrow streets and onto the I-5 interstate to Bellingham. We arrive at Squalicum Harbor to find frost on the docks and the sparkle of new snow on the distant slopes of Mount Baker. The marina where we spent two years as liveaboards on a twenty-seven-foot sailboat is nearly deserted. Even in a place where residents pride themselves on enduring the persistent winter monsoons, few boaters venture out in the middle of March. Normally, we wouldn't, either. But there's no time to waste if we have any hope of reaching the Arctic before winter.

With dry bags and gear strewn across the lawn, we unload our boats and begin to pore over the cryptic manufacturer directions for how to assemble a pair of oarlocks. We've forgotten whether green refers to port or starboard, one of the most basic facts of seamanship, and we stare blankly at the color-coded parts. "Fifty-fifty chance, right?" Pat asks and begins on one side. I grab the other oar. By the time we finish, the oars are attached securely but point backward like wounded birds with awkward, drooping wings. Our first failure.

After we swap the oars, we take a break for late-afternoon

sandwiches and I dig out the message a friend had left me as a send-off. *Remember to keep hands relaxed on the oars, straight back, pivot at the hips, and let your legs do the work. Catch...push with the legs...swing back...clean hands in and down to the finish...arms away...swing up...and slow creep up the slide to the next catch. Keep 2–3 beats up the slide to 1 beat pulling back and you'll glide along no worries.* Pat and I laugh drily about the fact that we don't know what a "catch" is, but we also pay attention. Standing in jeans and sneakers on a patch of green grass, we need any help we can get.

Instead of training for what will be a major endurance event, we've spent the last several weeks eating poorly, sleeping little, and forgoing exercise entirely. I can only hope that our bodies hold some of the same innate energetic capabilities that many migratory birds do. To prepare for days or weeks of travel, birds don't fly sprints on their winter territories or lift weights at the local gym. Instead, they stay put and consume as much food as possible, in many cases doubling their body weight in just a few weeks. "Eating like a bird" is clearly a misnomer. Geese graze on nitrogen-rich grasses or gorge themselves on corn. Sandpipers stock up on mollusks. Finches chow seeds. The same rate of weight gain in humans is almost unthinkable. So well adapted are birds to the rigors of migration that even their organs are flexible; in the great knot, a wading bird that flies between Siberia and Australia, nearly every body part not required for flight—from intestines to liver, spleen to skin—shrinks during migration. As modern humans, we're starting with a distinct disadvantage: no matter how much we train or how far we must go, we can't survive on feast and famine. If we're lucky, Pat and I might be able to transform some of our love handles to energy the way birds do.

Our fitness levels aren't the only problem. In all of our

frantic preparations for the trip, we had neglected perhaps the most important one of all—learning to row. Our total combined experience is a quick jaunt in a friend's creaky aluminum dinghy across a protected cove, and a lazy afternoon of fishing in a borrowed raft. Our initial plan had been to acquire rowboats long before we left so we could teach ourselves to row and troubleshoot any potential problems. We were committed to rowing—not only would it be more efficient than kayaking, but it would allow us to keep our legs strong for the hundreds of miles of skiing and hiking to follow. However, the fact that Pat had to build our boats, which he barely managed to finish in time, and that most water in Alaska was frozen solid when we'd left, made practicing impossible.

Perched next to the much larger skiffs and sailboats in the marina, our rowboats look fragile, more art than function, their shiny paint mocked by the dented steel and stained fiberglass of the Squalicum fleet. When I lift one end of my boat to help Pat carry it to the dock, the heavy, awkward load strains my back and pinches my knuckles. Each boat weighs nearly one hundred pounds, and we must haul them, plus our food and gear, up and down the beach at each campsite. I have no idea how I'll manage to carry the boats over slippery rocks and cobbles. I have no idea how I'll do any of it.

We lower my boat into the water for the very first time and watch anxiously for leaks, running our hands along the glossy cockpits, testing the insides of the hatches for moisture. Pat helps me climb in and pushes me away from the dock so my outriggers are free. It takes a dozen tries before I manage anything resembling a stroke. The oars have a mind of their own, and my seat slides up and down the length of the boat without waiting for my direction. My neck immediately begins to protest the fact that I will have to look over my shoulder for the next thousand-plus miles—facing backward is one of the inherent downsides of rowing. When Pat gets into his boat, he looks as awkward as I feel. Thankfully, the docks are empty of spectators.

That night, after finishing our packing, we stay in a friend's boatbuilding shop near the marina. As I unroll our sleeping pads, Pat calls to me from the basement. I walk downstairs and see him hunched in the corner, peering at a wooden box with his headlamp.

"It's *Sprucey*," he says with a grin.

Inside the box is the spruce canoe that shaped our relationship. Pat brought the canoe down to Bellingham by ferry after convincing his college advisor that a museum exhibit featuring the canoe was worthy of fine arts credit. After the exhibit, he stored the boat in our friend's basement, where it still sits. All these years later, it's fitting for *Sprucey* to see us off. That earlier canoe-building trip, almost a decade ago, doesn't come close to rivaling the present one in terms of miles, ambition, or sheer scale. The distance we covered then would hardly register on the map of our upcoming journey. But what we gained had nothing to do with mileage. Before the trip, our togetherness was tentative, a relationship still searching for traction. Our time on the river changed that. As I lean against Pat in the dark, he unscrews one corner of the plywood lid. The umber bark, stitched with yellow spruce roots, stares back at us.

"I guess this is what it means to come full circle," I say.

After we go to bed, I thrash around in my sleeping bag, dozing off only long enough to wake again with a start. My thoughts bounce between fears of capsizing in big waves to anxieties about the growing list of things I have forgotten to pack. Pat, lying on the floor next to me in his own sleeping bag, is sound asleep. He has put his worries aside in favor of rest, which we both desperately need. *It will be easier in the morning, when we're fresh,* he told me. It's a gift to be able to suspend doubt long enough to believe in the impossible. A gift I'm currently lacking.

The next morning, I stumble out of bed and across the street for coffee, craving nothing more than a lazy day of reading inside a warm café. But instead I pull on several layers of synthetic and wool clothing, don rain gear, and brace myself for what lies ahead. It's March 17, St. Patrick's Day. Beyond just shamrocks and green leprechauns, it's a celebration of the patron saint after whom Pat was named. I peer out from under my hood at the rain and wish for a bit of Irish Catholic luck today.

Two friends from Bellingham meet us at the marina and help us carry our boats to the dock. Before loading our brightly colored stack of dry bags, Pat notices that the bolts on the oarlocks are loose, threatening to slip into the water and disappear. He makes a hurried trip to the hardware store for replacements as I shiver next to my friends in the rain and watch a flock of harlequin ducks bob in the lee of the breakwater. Chop is beginning to form on the water. When Pat returns, we secure the bolts in silence and step over the large outriggers into our boats.

Before I push away from the dock, I pause to exchange a grin with Pat. Despite the uncertainties and the stresses of the previous weeks, it's finally time to launch! If nothing else, we've made it this far. We pose for a photo before leaving the comforts and burdens of our regular lives behind.

As soon as I begin rowing, I feel naked in the open cockpit, vulnerable to the wind and rain. Unlike a kayak's seat, the rowboat's sliding seat is perched high above the waterline, and there's no spray skirt to keep the water out of my lap. The oars are nearly ten feet long and unwieldy as I lift them up to return to the top of each stroke. To make the blades move through the water in synchrony, my right hand must pass over my left. It's awkward and I bump my thumbs nearly every time. I try to remember my friend's message about the catch and the beats. I know only that my beat is completely off. I let go of one oar handle to wave at our friends and it hits me in the chin. When I look over at Pat, I notice the fine creases around his eyes are etched deeper than usual.

By the time we reach Horseshoe Island an hour later, the cold March rain has turned to hail. I step out of my boat and try to shake the cramps from my arms and legs, made worse by tense overgripping and exaggerated strokes. Small, round blisters have started to rise on my palms, and my thumb is darkening to purple where I have repeatedly smashed it between the handles of the oars. As I stretch and shimmy, Pat does his own version of calisthenics, hopping from one leg to the other and pummeling his butt cheeks with his fists.

The restlessness I felt before we left Anchorage has been replaced by a growing dread. Unlike the godwits and gulls that inspired me as I sat in my office cubicle, I have no wings, no magic carpet that will carry me north. Neither my body nor my mind feels ready, and I'm rowing a boat that refuses to obey my commands. I had thought foolishly that *Zugunruhe* would be enough. I had imagined that the urge to move would take hold as soon as we hit the water. Instead, I've already started to question the reasons for this journey. The enormity of our goal is dizzying.

Two hours later, we reach the Strait of Georgia, where the shelter of Lummi Island gives way to miles of exposed rowing. As soon as we nose out into the channel, a strong southeasterly wind shoves us from the side and waves threaten to wrest my oars from my hands. A mew gull catapults past my head, leaving a stream of white guano on the bow of my boat. As a child, I'd been told that getting pooped on was a sign of good luck, but after years of working with birds, I know the truth. There's nothing lucky about it.

"It's too rough, we need to go back," I shout to Pat. I resist the ocean's attempt to pry me from my seat each time I dig an oar into a wave and slowly fight my way back to the lee of the island.

As we scan the shoreline for a place to camp, we see only backyards and an unruly assortment of gardens and sheds and old trucks. A tangle of seaweed and sticks is nestled against the base of the bluff, evidence that this narrow strip of beach will flood at the next high tide.

"Great, what do we do here?" Pat mutters.

"Ask if we can camp in someone's yard," I answer, refusing to consider the prospect of heading back out into the waves. Pulling our boats ashore, I run up the steep bank to knock on the door of one of the houses. No one answers there, or at the next two I try. Taking absence for tacit approval, we choose a spot hidden by a thicket of blackberry brambles, picking our way around old pieces of fencing and barbed wire, and anchor our tent fly to a cracked four-wheeler tire.

It's far from an idyllic beginning, but we're too tired to care. We fall asleep at dusk and wake to the sound of a car door slamming in the dark. No one visits until early the next morning, when a fat brown Labrador retriever wanders down to pee on a tent stake. Sneaking away just after dawn, we cook breakfast on the wet sand of the beach and launch toward Georgia Strait again. This time, the water is as still as a lake. It's a blessing we happily accept. With only seven miles of rowing behind us and nearly twelve hundred to go, we need every break the weather might offer.

* * *

Over the following days, as we pass through protected waters around the San Juan and Gulf Islands, I slowly learn to move with the water rather than against it. Though stiff and sore each night, my muscles loosen to the rhythmic motion of rowing. Eventually my hands feather the oars intuitively and I can finally hold a straight course. I see gulls and ducks cruising past and no longer curse them for mocking me with their easy wingbeats and sure purpose. Still, the unsettled spring weather sends us careening frequently to shore in search of shelter. We wash up on beach after beach to find most summer homes empty this early in the year, with no one around to mind a couple of bedraggled boaters camping on their lawn. Occasionally we pitch our tent unannounced only to realize later that we have slept in someone's yard who *is* home or have blocked the path of a neighbor's daily morning walk. Instead of angry words or the barrel of a shotgun, we almost always face good-natured hospitality.

One man apologizes profusely when he finds us in our tent at 7 a.m., sheepishly occupying a dormant flower bed. "If I'd known you were here I would have invited you up for hot drinks!" he tells us. At another beach, we don dry suits while a recent retiree from the mainland checks the forecast and calls ahead to his friends to keep a lookout for us. An hour later, we look up to smiling shouts of "Bellingham!" and a waving flag. These first small acts of kindness from strangers buoy us through the rain and wind.

Shortly after leaving the port town of Nanaimo, we are forced back by big chop and gusty winds and find ourselves hauled out on a paved path at a public dog-walking park. Trying unsuccessfully to be inconspicuous, we hunker down near our boats in the driving rain and wait for darkness, when we can pitch our tent unnoticed. Before long, an elderly gentleman with a strong Norwegian accent approaches us in the bushes and invites us to his house. "Too bad my wife isn't home. She'd know just how to take care of you. I could heat up a frozen pizza," he says, "and put your clothes in the dryer." He will sound surprised when I call his number several months later to thank him, as though an overnight invitation is nothing more than a simple courtesy.

But friendly offers of food or a place to stay are almost always followed by surprise when we explain our circumstances. We face blank stares as people look at our rowboats, look at us, and ask where we're going. *Alaska?! In those boats?* Again and again, I nod and force a smile, pretending that my own doubt doesn't mirror their own.

STORM DAYS

I used to think I loved the ocean. Today I hate it. Swells maw open at irregular angles and threaten to swallow me. The waves have teeth. Gusts wash over me with the cold, sucking air of a passing train, forcing me to brace rather than row. I concentrate on keeping my hands low on the upstroke to avoid the jarring force of the water grabbing my oar. I try to think about anything except the fact that I *can't* get out right now.

Pat rows his boat thirty feet away, riding the sea like a new jockey. His eyes are fixed; his jaw is clenched. With each set his bow plunges into the trough of a wave before rising again. Normally he would be checking in with me every few moments, offering a glance, a hand gesture. But for now he is consumed by the water.

The storm that has been threatening to blow up for the past several days has finally arrived, announcing its presence suddenly and offering us little time to find protection. It's the third gale in less than two weeks, and this one promises to be the worst. Since leaving Bellingham, we have come to accept that this is the season of storms and driving rain and frost on our boats each morning. We have worked to relax into the steady cadence of pull-glide, pull-glide, to exhale with the creaking of the oars. Right now those lessons seem abstract and distant. The fear burning in the back of my throat does not.

Please, let me be anywhere but here. Head down, peering out from under a deluge of rain and sleet, I'm holding on one second, fighting for control the next. Again and again the waves threaten to leverage me into the water. With each stroke, the hard plastic handles of my oars collide and graze my knuckles. "I don't know if I can do this!" I shout my panic into the wind. Briefly, the sound of a human voice, even my own, mewling and high-pitched, makes me feel less at the mercy of the sea. My boat rolls and lurches and I channel terror into rage—at the waves, at the wind, at this stupid trip.

Early this morning, as we listened to the forecast on the VHF radio from the warmth of our sleeping bags, we debated whether we would have enough time to make it past the next town, Campbell River, before being pinned down by the impending blow. The computerized voice warning of thirty-five-knot winds and six-foot seas, predicted to hit by evening, made me stiffen against Pat. "Well?" I prompted after we turned off the radio. With his usual optimism, he said he thought we'd be fine, that we could cover a few miles before we would have to find another campsite. In return, I mumbled something about spring being too stormy for rowing, even as I began to pack up the tent. We both knew that sitting out the weather on a perfectly calm morning wouldn't get us any closer to the Arctic.

For the first two hours, my worry seemed silly. We glided through water so still and clear that our eyes could trace the stalky arms of bull kelp down to their anchors on the sea floor. Crows and ravens picked at foamy yellow mats of herring eggs along the high-tide line. We spotted our first humpback whale, feeding in the shallows. The whale's rounded, silvery back rose from the water with a steamy exhale. Nothing in its graceful movements hinted at the journey it had just completed or the desperate hunger it must have felt only days or weeks ago. Migrating farther than any other mammal on the planet, humpback whales swim more than three thousand miles each way between Hawaii and British Columbia or Alaska each year. When they return to northern waters, they feast on the smorgasbord of silvery herring and tiny crustaceans, gulping giant mouthfuls of plankton and krill.

As I watched the whale surface, I remembered why being here mattered. Traveling north in the springtime is among the most primal of acts. With salt from sweat and seawater dried on my lips, I felt part cetacean myself. But that was before the wind began to ripple across the water, before we realized how few safe landings this coastline offered.

Campbell River, a community of thirty-five thousand people nestled against Vancouver Island's eastern coastline, begins to emerge from the fog. Although we're less than three miles from the edge of town, there's no mistaking our aloneness. In theory, we're capable of self-rescue. Under most conditions, if I capsized, I could right my own boat and climb back in. Today, in a storm a half mile offshore, I'd be lucky to grab my boat without getting knocked in the head by an oar. I'd be even luckier to hold on long enough for the possibility of help to arrive. In this zone of big surf and forty-five-degree water, there's only one option—we simply can't capsize.

Given their size, our eighteen-foot rowboats have proven to be seaworthy, even powered as they are by novices. With the configuration of the sliding seat, we can direct all of the energy from our arms, backs, and legs into the long oars, which act as outriggers. The boats are much wider than typical racing sculls, nearly three feet at the hips. Their hulls have just the right shape, offering efficiency without compromising stability.

But no matter how well designed, eighteen feet is only eighteen feet. Today, our boats feel tiny in the waves.

I notice a pair of headlights flickering down the coast. Lights from buildings and houses are bright against the gray sky, and I imagine what it feels like to be inside. Warm. Quiet. Safe. For an instant, I escape to a previous version of myself—a young girl curled up with a book and a cup of tea, reading about someone else's crazy adventures. But the storm continues to howl and we need to act. The tide is falling quickly and the rocky shoreline churns with breaking surf. When a wave splashes in my lap, cold water seeping between my thighs, I fight the urge to call aloud for help. I pull harder on the oars with each stroke and glance over my shoulder just long enough to confirm that Pat is still nearby. There's nothing we can do for each other right now, except to stay upright in our own boats.

Eventually, a small beach appears through the mist. I want to be there, now. I turn and see Pat already nosing his boat slightly toward shore. We make it only a hundred yards before spray ricochets off of a submerged reef in front of us. The landing might be safe, but getting there clearly isn't; the fact is that we're better off out here in deeper water. If we choose poorly, our boats could be crushed in a second, sending us somersaulting into the frothy mess. Pat uses his arms to signal what I already know. We'll have to keep looking.

After twenty more minutes of wrestling the oars, a boat landing with a simple breakwater and launch ramp comes into view. I shout to get Pat's attention. When he stares blankly at me in response, I let go of one of my oars to point and give the thumbs-up. Pat returns the gesture and we angle toward shore again. As we near the breakwater, I pause long enough to swallow my pride. Pat is no more suited to the landing than I am. Except he isn't afraid to try. When

I gesture for him to go first, he doesn't protest. He studies the water carefully before bursting into motion, rowing hard against the current of the outgoing tide. Once he passes into the lee of the breakwater the waves relent enough for him to jump out. In waist-deep water he drags his boat partially out of the surf and waits for me to follow. I take a deep breath, watch for a break in the sets, and pull with everything I've got. As soon as I pass the breakwater, he's there to catch my bow. Together we muscle the boats out of the water and collapse in a soggy heap on the gravel.

"We did it," Pat says as he reaches over to take my gloved hand in his.

"Barely," I reply.

I am content to lie here forever. At least until my toes start to complain about being cold. Until the beach begins to feel less like soft sand and more like lumpy gravel. Only after discomfort fully edges out the lingering adrenaline do we stand up to take inventory of our surroundings. When the storm first threatened, I envisioned snuggling in our sleeping bags on a quiet beach with big trees for a backdrop. Pavement and cars weren't part of this equation. We have landed at a public boat launch surrounded by houses. Across the street are a gas station and a motel with a neon red vacancy sign. We had planned to sleep in our tent, as we would nearly every other night for the next six months, but the temptation of showers and a place to dry our clothes is too great.

"What are the chances of washing up within one hundred feet of a motel?" I ask. Without waiting for Pat's response, I jog across the street and push open the scuffed office door, bells swinging overhead. The man at the front desk takes one look at my dripping hat and hair, bulky life jacket, and bright orange dry suit and asks me in a thick Australian accent only half jokingly if I've lost my boat. When he tells me that they have discounted rooms for the winter, I reach across the counter to hug him. He pulls away and hands me a room key, muttering, "Fooking crazies around here."

The motel room offers an abrupt transition from bobbing in the surf to sprawling on a flowered bedspread, and we waste no time settling in. Under steaming water, I whittle a small bar of complimentary soap down to almost nothing as I scrub my body and my hair, using what's left to wash our clothes in the bathtub. Lying pink-faced in socks and a towel on the king-sized bed, I could be in a luxury resort in the tropics rather than a dumpy motel on the battered North Pacific coast. In our extravagance, we order an extra-large pizza and devour it easily in fifteen minutes. Still hungry, I pull rain gear over my nakedness and make a visit to the gas station next door. For eleven dollars, I buy them out of super-size Reese's Cups. When I return, I curl up next to Pat and sip the tea he's heated in the microwave.

We stay up much later than we should, not falling asleep until we've finished channel surfing through a series of sensational news stories that feel like they belong in a different reality. When I wake in the middle of the night, Pat and I are nested closely together as though we're still in our tiny tent. I hear the strange sound of canned laughter and roll over to click off the television remote.

For the first two days, as the storm continues to rage, our unplanned stopover feels more like a vacation than a stranding. As we hunker down in the motel room and wait for the weather to improve, we repair gear, catch up on journal entries, and call family and friends. One evening, Pat dictates the first postcard to his grandparents, a habit that will soon become part of our town routine. I laugh as I write. "We're stuck in a town called Campbell River. Things could certainly be worse. The ocean is rough but the bed I slept in last night is awfully soft. Hope you two are doing well."

But by the third 5 a.m. wake-up to a shrill alarm clock, and with a continued forecast of high winds and big seas, the novelty of waiting has worn off. We stop calling our families to check in—weathering a storm in a motel hardly seems an adventure worth recounting. We pore over maps and grow increasingly anxious to swap the stale odors of used bedding and stained carpet for the briny smells of the sea. As rain streams down the windows, I tally how far behind schedule this stopover is pushing us. We counted on being able to row more than twenty miles a day. So far, we have barely made ten. And now, three more days of averaging zero. The math is simple: Every day that we're stuck waiting reduces our chance of success. Only one hundred and twenty miles into a four-thousand-mile journey, I'm feeling less like a bold adventurer and more like a failure.

While we were rowing, our needs were pared down to the essentials: eat, sleep, move. We rose at dawn, rowed until we no longer could, and curled gratefully in our sleeping bags each night. I was learning to let my body lead, with the ocean as my guide. Now that we have stopped moving, my mind has taken over again and I find my thoughts drifting back to what I left behind.

I had applied, shortly before leaving, for a postdoctoral fellowship to work with a well-respected wildlife disease ecologist and a renowned mathematician on the same beak deformities I've studied for the past five years. When I submitted my application to the National Science Foundation, I did it less for love of the subject—mathematical modeling of the avian beak—than for the fact that it seemed like the right thing to do. Even now, the promise of working with high-caliber scientists at Princeton is appealing, though moving to

New Jersey and studying beaks again is not. The award won't be announced for several more weeks. Another job I had applied for with a federal agency in Anchorage is also pending. Each of these prospects leaves me tense and confused. Although I know I can't bob around in the ocean endlessly, I'm no longer sure what it is I want.

I check my old university e-mail account and see the names of several fellow graduate students listed as the recipients of conservation awards and funding grants. As much as I was ready for a break from my studies, I can't help but feel a twinge of jealousy. Others seem so driven, so certain about their paths. Instead, I'm sitting in a motel room on Vancouver Island hoping that a floundering journey to the Arctic will offer some clarity.

To make a trip like this possible, we've had to give up more than just professional commitments. My sister is due with her first baby in less than a month and I won't be there for the birth. My dad's health is steadily worsening, and I've recently done little to help or even to acknowledge his situation. We've used most of our savings to purchase supplies and organize logistics. As I lie awake in the hotel room at night, restless from too much junk food and inactivity, I feel like a thirty-three-year-old kid still waiting anxiously for the answers that are supposed to come with growing up.

Finally, on the fifth day, the steady drone of the wind against the motel's cheap windows eases and the quiet emanates like a giant sigh of relief. We don't have to listen to the weather radio to know when it is time to go. Leaving is easy. Our bags have been packed for days. We grab one last coffee and sugary pastry from the gas station next door, tiptoe out in the dark, and load our boats in the gently lapping water.

Within an hour, our rowing companions number in the

thousands. Scoters congregate in rafts so large that the round white patches on the backs of their heads blur into a pointillist painting as they dive in synchrony. Dozens of sea lions cavort near our boats. Sleek and graceful underwater, they explode from the surface like waves crashing on a reef, tossing their brown bodies through the air in a show of raw power. Herring school in silvery masses below tornadoes of foraging gulls. I catch a glimpse of a rufous hummingbird as it passes overhead, its small body backlit by the pink sky of early-morning fog.

In the days before the storm, we saw dozens of hummingbirds, often crossing over large expanses of open water. Several came to investigate the bright red deck of my rowboat—a giant faux flower with no promise of nectar. I decided then to christen my boat *Rufous* in honor of these tiny travelers, and in fond memory of my conservation biology professor. Less than two years after I graduated from college, I received news that Bill had died following a short battle with leukemia. We had stayed in touch by e-mail, but I never properly thanked him for what he had given me: a new way of seeing the world.

As the hummingbird flies past, I recall a photo of Bill's gnarled hands, showing a silver leg band so minuscule it fit within the smallest whorl of his fingerprint. At just over three grams, a rufous hummingbird could be sent across the country for the cost of a postcard stamp. But as Bill was quick to point out, size doesn't always scale equally with toughness; proportionally, these birds make the longest known avian migration on the planet. On miniature wings that beat two hundred times a minute, they travel clockwise, up the Pacific Coast in spring and down the spine of the Rocky Mountains in late summer. Their wintering grounds in Mexico are separated from their Alaskan breeding grounds by nearly 80 million body lengths. For us, this would be equivalent to traveling eighty-three thousand miles—more than three times around the earth and just shy of the distance we might hope to walk in an entire lifetime. That's only one way.

That first summer with the kittiwakes, I found a rufous hummingbird nestled in a depression in the ground, its still form pressed against a bed of snow and dry brown leaves. It lay unmoving as I reached down to stroke its iridescent green head. A casualty of a late-spring storm, the last of its reserves sapped by the recent snow, I figured. But when I picked up the bird and cradled it in my hands, its fragile body slowly stirred to life. It was cold, not dead. Each day, rufous hummingbirds must eat three times their body weight to power their flight. At night, they enter a state of torpor that allows their core temperature to drop and heart rate to slow, just as chickadees do in the dead of winter.

These birds survive by only the narrowest of margins, but they act like they own the world. Watch one at a feeder for a few minutes and you'll see a tiny bird with a big attitude. They thrust their swordlike beaks at potential intruders, including birds twice their size, and leave no question as to who has claimed rights to the nectar. Their pluck has reportedly scared off even chipmunks and eagles. Bill told me stories about their grit, the way they fought against their human captors like they had every chance of winning.

As the wind rattled our motel room windows, I had wondered how the hummingbirds, arriving before any blossoms or buds had emerged, fared in the storm. Now, with each pull of my oars, I remind myself that it's naive to question the resilience of a bird that weighs little more than a penny and migrates twice the distance we plan to row in less than a quarter of the time.

Heading north into the crisp morning air, our own bodies well rested and well fed, anything seems possible.

CHASING TIDES

After almost a month on the water, I'm suspended in darkness, my boat slicing the border between a black sky and a blacker sea. Sea lions belch and murres moan; voices that are familiar by day become disembodied and eerie by night. The smallest waves startle me in this strange world of shadow and sound as I search for the neon illumination of phosphorescence beneath my boat, hoping for some definition in the inky water. When my oar collides with a dead surf scoter floating facedown, the white patch on the back of its head stares up at me like an unblinking eye. Pat is only thirty feet away, but I long for the comfort of closeness. Because of the broad reach of the oars, it's impossible to row in close tandem as we would in kayaks. I can't make out the silhouette of his torso; instead, his oar blades flash white and ghostly against the glow of his headlamp. It seems crazy to continue like this, in the dark, but we have no choice. We can't find a beach to land our boats. Every cove is swollen to its edges and cliffs stare blankly at us as we drift by.

We're just a few days past the spring equinox and the solar system has aligned, quite literally, creating an impressive gravitational pull. Spring tides are often large, but this cycle has trumped our expectations. Over the past several days, record tides have sent us clambering up steep banks to find dry campsites and forced us to haul our gear across miles of kelp-strewn beaches. Water found its way to places that are normally dry, lapping at tree roots, creeping to within inches of our sleeping bags, and leaving the tent's guylines dangling in brackish water. Hours later, the falling tide exposed mussel beds that rarely see light. Though even at their most extreme, local tides are nowhere near the world's largest—which can vary more than fifty feet, as they do in Nova Scotia's Bay of Fundy—the scale is relative. It's the biggest tidal range this region has seen in nearly ten years.

Our tide book reports that tonight will bring the highest tide of the year. The little book's modest design masks the complex forces that shape these numbers: the moon spinning around the earth, the earth around the sun, gravity that varies across the surface of the ocean. It's an interplanetary dance that requires sophisticated physics to understand. But, behind the calculations, there exists a basic truth: our watery blue earth is a small piece of something much larger than we will ever comprehend. For us, these monster tides also bring another set of lessons about the sea, and about life.

Two nights ago, we pulled into a narrow cove under the falling darkness, hauled our boats above the tide line, and climbed into our sleeping bags. Several hours later, I woke to the sound of wood on wood, a hollow thudding that nudged me awake. As I startled from sleep, it took me a moment to locate myself in the dark forest, wedged between salmonberry thickets and the shadows of stalky hemlocks. Behind Pat's gentle snores, logs banged against each other in the surf. Suddenly the sound registered in my mind and I shook Pat's shoulder as I began frantically unzipping the tent.

"Wake up! I think the boats are floating!"

We scrambled to dress, crammed our feet into our boots, and climbed out of the tangle of bushes. The water had risen to meet the mossy forest floor and all the driftwood in the cove was afloat. Under the narrow glow of headlamps, we picked our way down a jumble of logs near the shoreline. I slipped on a slimy cedar and splashed to my waist in the cold water.

Following behind Pat, bleary-eyed and dripping, I prepared myself for the worst—our rowboats crushed among the drift-wood, buried beneath logs larger in diameter than I stood tall. As we rounded the bend, I stared at the place where our boats had been. I blinked and squinted into the darkness, try-ing to make sense of what I was seeing.

"Pat, I can't believe it," I shrieked as he stared at the same startling scene. Only two feet lower and our boats would have become flotsam. Instead, they were perched on the sin-gle stationary log in the sloshing fishbowl that was once a beach. The high tide had barely missed picking them up. We tiptoed along the water's edge to inspect our boats, then sat down in silence for several minutes, pressing our hips and shoulders together against the damp night air.

Back in the tent, clothes changed, my mind began to race. Earlier in the evening, I hadn't been able to shake a nagging feeling that I should check in with my family, but our satel-lite phone refused to pick up a signal among the large trees. This is one of the hazards of straying beyond the reach of communication: the worry that comes with not knowing. I cycled through a hundred different disasters that could have befallen the people I love—car wrecks, heart attacks, ski ac-cidents. My sister's due date was also less than two weeks away and I was anxious to hear her news.

I had learned about my sister's pregnancy just as we began

preparing for our trip. She was three and a half years my junior, and her announcement reminded me that decisions about having children couldn't wait forever. At the time, this had seemed like yet another reason to go. I knew I needed to leave while I still could. But out among trees and water, I felt acutely sad about not being there to share in this major passage of her life, and mine. We had become especially close after I'd left for college, the years between us shrinking as we grew older. With labor imminent, I thought constantly about her health, and the baby's.

Pat woke occasionally to my stirrings, stroking my forehead and reminding me that our boats were fine and we were safe. I lay awake for hours, increasingly anxious as the first light began to brighten the tent.

In the morning, we launched into a headwind and building chop. After ducking into a cove to don dry suits, we reached a protected channel and began to head west, toward the island community of Bella Bella. Several hours later, at the barely floating public docks, I pulled my boat up next to empty soda bottles and a shopping bag full of clam shells. Before I had taken off my lifejacket, I dug out the satellite phone. Under the curious stares of several local children, I heard my sister's voice on a message. I held my breath as I waited for the news—after an epic two-day labor she had delivered a healthy baby boy.

As soon as the message ended, I dialed Ashley's number. Her husband answered. "She's sleeping," Scott said. "She needs the rest but I know she'll want to talk to you. We named him Cormac. And Willem, after your dad. He's perfect."

When I hung up, reeling with the news of my nephew, I stood dumbly at the docks while skiffs in every state of disrepair buzzed in and out of the harbor. We were conspicuous with our still-shiny rowboats and colorful gear in this modest

First Nations community, but besides a few kids playing near the water's edge, no one gave us much more than a glance. The only exception was a weathered, pale-faced man who stared at me as he chain-smoked in his shiny white SUV. After ten minutes of his awkward attention, I walked over to the car window and said hello, tense before I'd opened my mouth.

"I'm Rick." He stuck his hand out the window to shake mine, our equally callused hands a match.

"Caroline," I replied.

"Where you staying?" he asked, part invitation, part threat.

"I'm here with my *husband*," I told him pointedly, "and we're not sure yet. We just pulled in."

"I have an extra bedroom in my house. It's on the water, you can leave your boats there." His face transformed from scowl to smile as he continued. "My wife works at the hospital. She'll take good care of you. Here's the address if you decide you want to come."

He pressed a scrap of paper into my hand, and pointed south along the shoreline. "It's a quarter mile down the beach, a yellow duplex. You can't miss it." At first I wanted nothing to do with this man or his strange offer. Only after he drove away, tossing his cigarette out the window, did I reconsider. It *would* be nice to have a shower. And there was nowhere obvious to camp. I dreaded the thought of fighting with the tides again tonight. Most of all, I really wanted to talk to my sister before we left town.

"I guess there's no harm in checking it out," Pat said when I recounted the conversation.

Smoking on the back porch of a tidy duplex, Rick was waiting as we pulled up an hour later in our boats. I softened slightly when I saw two hummingbird feeders hanging above his head. As promised, his wife, Katherine, hustled down to the beach to greet us.

"Rick's back is bad but I can help you unload. Come in, have a shower. We'll throw your clothes in the washer. You must be ready for a soft bed!" She was all charm and chattiness, drawing me in as easily as her husband had pushed me away.

While we scrubbed our bodies in their clean tub, Rick began to prepare a feast of locally harvested seafood. As the prawns turned from pink to red in a sizzle of butter and garlic, he tossed a few choice pieces of meat to a fluffy white sausage of a dog. I could already see the kindness beneath his gruff facade. When we sat down at the table, I mentioned that my sister just had a baby. "Oh, you'd better call her right away!" Katherine said. "Kids weren't in the cards for me. But I wish they had been." We ate in silence for a minute before she continued. "There's no substitute for family. Don't ever take your sister for granted."

In a brief whirlwind of a story, she told us how she had treated an ailing woman over a period of several years at a health care clinic in the small Vancouver Island community where she had been raised. Shortly before meeting her patient, Katherine had learned of her own adoption, and of sixteen biological siblings she'd never met. Katherine's patient had also been adopted, and they chatted frequently about how much they each wished they had known their families. Though they spent time together only at the clinic, Katherine explained that the connection ran much deeper than anything she'd experienced before. "I saw plenty of sickness and death, but for some reason the fact that I knew she wasn't getting any better made me sad. Much sadder than I could understand at the time. By the time I realized why, she was gone."

Months after her patient had died, no closer to finding her biological roots, Katherine learned a shocking fact. The woman was her sister. "We both came that close to having family. But we missed the chance. Call your sister," she insisted again. "Dinner will wait."

So I did. When Ashley answered, her voice, hoarse with sleep, was strong and steady. There was no bitterness when she explained how nothing had happened as she'd planned. She started in a birth center with a midwife but ended up in the hospital with IVs and monitors after her labor failed to progress. "It wasn't what I expected, but I don't think there's anything more I could hope for. Cormac's heart rate kept dropping, but he's OK now."

When I asked her if childbirth was really as crazy as it seems, she said only, "Crazier. And now I have a son. Somehow that's the craziest part of all." From behind her words emanated a stillness and composure I'd never noticed before. It was more than simple exhaustion. I was hearing the voice of a mother. Ashley and I filled our conversation with several long silences, not for lack of things to say but because sometimes words are unnecessary. After one of the pauses, I told her about Katherine's story, about the fact that she only came to know her sister through death. "Imagine, sis, if we never knew we existed to each other. Until it was too late."

Now, with Bella Bella and my sister's news thirty-five miles behind us, I weave between forested islets, navigating by both sight and sound. Straining to see, I search for the shiny heads of seals as they rise with an exhalation and disappear again with a splash. As the moon's huge yellow orb rises above the water, I force myself to drop my shoulders, easing the tension that binds my muscles and begins to creep up my neck. I remind myself that I know how to row. It's dark, but I can do this.

My body rocks forward and back as my hips slide up and down the seat with each stroke. Pull, glide. Pull, glide. Pull, glide. The oars answer with a gentle creak-creak, creak-creak, creak-creak—echoing a simple, soothing cadence, like wing-beats through the dark night. Repetition unravels my frayed nerves. Movement brings stillness. Mind follows body. The current has shifted in our favor and we cruise along easily in the night. I am lulled into complacency, then awe, as soft, ethereal light reflects on the fjord's narrow walls.

Channeling the calm Ashley exuded through the phone lines, I recall where we had been while she was in the throes of labor. We had covered nearly forty miles that day, pushing our muscles beyond comfort, then beyond discomfort. Blistered hands scoured with salt water as rain turned to sleet, I dug a little deeper with each stroke, asking the impossible of my body, while Ashley asked more of hers. Later, we bumbled around in the dark in search of our boats, wet and cold and scared, as she spent the first moments with her new son.

As my thoughts drift into the night, I begin to notice faint sounds coming from the sky and hold my oars still to listen. It's difficult to discern the precise source or direction of the voices, and I begin to wonder if fatigue and darkness are clouding my senses. But when I see a few small flecks silhouetted against the moon, I know we're not alone on our midnight journey. Many species of birds travel in the dead of night, flying across water and land. It's yet another improbable fact of migration. The reasons for nocturnal flight are many—greater opportunities to feed during the day; reduced turbulence; cooler temperatures that limit evaporative water loss; avoidance of predators. Birds also happen to be good at orienting themselves by the stars.

My oars slip into the black water and I remember Ashley's words just before we hung up the phone. "We're on such different paths right now, but in a way they're the same, too. Every day brings something new. Something we'll never forget." She's right. Just when the ocean has started to feel familiar, everything is new again.

Rowing in the dark through the biggest tide of the decade, there is no night but tonight, no time but now. Perhaps becoming a mother is much the same, every moment precious in its transience, each day unlike any other. My sister has just begun a journey with a newborn who will soon become a toddler. She will hear his first word. See his first step. She is learning what it means to create a family. For us in our rowboats, there are birds passing in the night. Yellow herring eggs that are present for only a few days each year. Tides that leave us humbled and reverent. A trip that can only happen once. Despite all that we have before us, I can't help but wonder what rewards a child brings that rowing through the dark cannot. Even in her cheery hospitality, Katherine's sadness was palpable at the dinner table. The absence of family, at least in the way she had wished, left a hole impossible to fill. It wasn't as simple as regret. It was something more. A missed opportunity. A connection that could have been.

Three hours pass before we reach an island with a narrow strip of sand. After hauling our boats out of the water, I rifle through our dry bags for late-night staples—a block of cheddar cheese crusted white with salt and a half-eaten bag of peanut M&Ms. We fall asleep with multicolored fingers, fragments of the last candy smeared red against the floor of the tent.

The next morning, we drag ourselves out of our sleeping bags long before we're ready. The beach that was nearly underwater when we landed in the dark has grown to ten times its former size. Its white sand entices us to spend a lazy morning relaxing in the sun. But as I pack up the stove, the birds quiet in the predawn mist, I remind myself that we must continue. We're in a land of rain and big trees. The Arctic is still impossibly distant. There are four mountain ranges, two seas, and a dozen rivers ahead of us. It's time to go.

In order to stay on schedule, we have to follow the ocean's clock, not our own. In many places along the Inside Passage the tides funnel their energy in and out of narrow channels, and the resulting currents move faster than we do. Ignoring this fact is an invitation to struggle upstream, each stroke delivering a view identical to the last. Sometimes the consequences are more dire. The tightest constrictions boil with whirlpools and rapids that threaten to swallow an unwise boater.

Today, our battle with the currents will require more strategy than gumption. We're headed toward Princess Royal Channel, a strikingly narrow, steep-sided feature that stretches for thirty-eight miles like an arrow through the surrounding mountains. Though the strength of the current is not dangerous, it's enough to keep us from making progress if we battle it, or to scoot us along at nearly twice the speed if it's in our favor. The trick of timing here is that the tide pushes water from both sides of the channel to its middle, where the opposing currents meet and then reverse flow. This means we have to plan our rowing schedule to coincide with the flood tide on the way in and the ebb tide on the way out, which is not as easy as it sounds, especially when favorable tides begin long before the sun has risen.

As we pull away from camp, a faint breeze whisks loose strands of hair into my eyes and I hear a tiny, familiar voice saying, "Here we go again...!" I wish that for once we could move through our day with no second-guessing, no need for plan B or C or D. But it's too much to ask of the fickle spring weather. Embedded in all of our careful calculations about currents and tides is the other major player—the wind, ready to negate any plans we might have. Within an hour, the current is strongly in our favor. The wind, however, is strongly opposed. I turn around in my seat to look down the length of the channel and catch a gust in the face. The two forces have collided into a slurry of whitecaps that are more frustrating than scary, but it's obvious we won't make it to our midpoint anytime soon. Pat shouts something that is lost to the wind. I gather what he is saying from his gestures, waving an arm toward the nearest shore.

On days like this, there's no point in fighting back. We decide to stop at one of the last protected bays while we can still find a campsite. We soon discover the remains of an old cannery and spend the afternoon imagining what this place must have been like a century earlier. While Pat picks through crumbling block walls, I call my sister on the satellite phone. When she answers, I hear my newborn nephew crying in the background. Pat walks over in time for me to press the phone against his ear. His squint is followed by a smile as he registers the mewl of an infant a thousand miles away. When a flock of snow geese passes overhead, I hold the phone to the sky for Ashley to hear. Through a spotty phone connection, we exchange these sound bites from our lives.

That night in the tent, my conversation with Pat spirals around the idea of having a baby. Parenthood is something we'd always assumed we would get to eventually. But the eventual is beginning to feel less and less distant. *What if we did?* we ask ourselves. *How would our lives have to change?* And then, *What if we didn't? Would we always wonder what we had missed?* Tonight, these questions have no answers. For now, I will pore over the photos that Ashley sends of her son changing daily. And I will tell her of the birds I see, the ocean's patterns, the way the wind feels when it blows across my bow.

RULES OF THE SEA

Shortly after crossing the unmarked border separating Canada from Alaska, we pull ashore at Cape Fox, a thin strip of sand buffered by an offshore island. At the leading edge of a 978-millibar low, the barometric pressure is dropping precipitously fast and the wind has begun to gust and swirl. The storm that is forecast to arrive this evening has been described on the radio as a "meteorological anomaly." Here, on the North Pacific coast, such a report can mean only one thing—a huge blow. Wedged between two fin-shaped beaches peppered with deer and marten tracks, our campsite faces the open waters of Dixon Entrance. This is one of the most exposed places on the entire Inside Passage; to our west lies almost six thousand miles of open ocean. It's a straight shot to Micronesia from here. Inside the tent, protected by a stand of moss-covered cedars, we fall asleep to the steady patter of rain.

The next morning, the VHF crackles with ocean buoy reports of seventy-mile-per-hour winds and twenty-five-foot seas as we snuggle deeper into our sleeping bags. Even the ferries have been canceled today. Under storm surge and big surf, our beach is beginning to collect flotsam—bull kelp torn from the sea bottom; a plastic water bottle; hundreds of brown and white feathers; the carcass of a tanner crab. Soon, there will be much more. Just months ago, a tsunami ravaged the Japanese coastline, sending the remains of entire cities into the sea. For those struck by the tragedy, much has been lost and nothing gained. But for beachcombers on this side of the Pacific, it means an unprecedented arrival of salvageable goods. Fishing boats, lumber, and glass buoys feature among the estimated million tons of debris that are predicted to travel east across thousands of miles of open ocean.

Some of the most popular public-interest stories have already made the news. A soccer ball landed on Middleton Island in the Gulf of Alaska and was later returned to the Japanese child who had lost it. A crated Harley-Davidson motorcycle washed up on a beach in the Haida Gwaii Islands of British Columbia. A Japanese hand-built skiff was retrieved, nearly intact, by an American boat builder. These reports relay the happy tales of items lost and found. More often, of course, the sea does not return what it takes. Each found treasure is accentuated by loss—of lives, of homes, of the security that comes with a belief that disasters happen to someone else.

Back in the tent, waiting out the storm-force winds, I fill the afternoon reading a book called *Deep Survival*. Though it's a narrative about horrific accidents and gore, the book suggests a logical unraveling of tragedies that, at first glance, seem to result from a convergence of unfortunate circumstances. The author attempts to explain who survives extreme situations and why. I would like to believe that we can reason ourselves out of danger, that smart choices and a solid constitution can get a person through anything. But I don't really buy his argument. I've spent enough time among skiers and climbers and paddlers to recognize that, too often, luck trumps skill. There is always a certain level of risk involved in negotiations with wild places and wild elements. Even those places that seem tame, or familiar, like the river where my dad lost his best friend, can claim our lives. The key is finding a balance—trying to determine whether the risk is worth the reward. For us, right now, the equation is simple. At the edge of a volatile and unforgiving ocean, waiting is our safety margin.

As I crawl out of the tent to pee, masses of birds zoom by. In just the minute or two I'm outside, a flock of swans passes high overhead, hundreds of small shorebirds cruise along the shoreline, and a stream of gulls soars above the island. They are all heading north, most toward Arctic breeding grounds. We know from satellite tracks that storms can be both helpful and harmful to migrating birds—enough wind and they make hundreds of miles with almost no effort; too much and they might not make it at all. Like us, they must play the odds.

There's growing evidence to suggest that birds can foretell the weather. From warblers to geese, most migratory species seem to know just when to launch. For instance, successful southbound flights of bar-tailed godwits—those ultra-marathoners of the sky—almost all coincided with favorable tailwinds. Exactly how birds gather the necessary information to make their decisions is still up for debate. Large-scale atmospheric weather patterns are somewhat predictable, and thousands of years of evolution have likely shaped avian behavior to take advantage of such systems. Birds rely on unique sensory abilities that make them akin to miniature avian weather stations. Receptors near the feather follicles of the wings act as a traveling anemometer, allowing birds to judge wind speed when they are on the ground or airspeed when they are in flight. They're also sensitive to pressure changes, and may perceive approaching storm systems just as a barometer does.

Even so, birds sometimes get it wrong. In a constant game

of roulette with the elements, they end up pummeled, punished, caught in the wind and waves. Reports of migrating birds downed by rain, hail, or snow pepper the scientific literature. One of the earliest published accounts of storm-related mass mortality dates back to April 2, 1881. A biologist sailing thirty miles south of the Mississippi River in a gale witnessed more than two dozen species of birds—ranging in estimates from "quite a number" to "abundant"—flailing, drowned, or washed up on the deck of his sloop. Incidentally, this observation also validated the previously unconfirmed suspicion that terrestrial birds migrate across the open water of the Gulf of Mexico. Twenty-five years later, a Lapland longspur tragedy struck several towns in Minnesota; 1.5 million birds perished in a single snowy March night. The small Arctic migrants, confused and weakened by the storm and perhaps attracted by the towns' artificial lights, crashed to the ground. The next morning, they were seen strewn across yards, sidewalks, and frozen lake surfaces, their brown and white bodies broken and battered. Although extreme weather events are clearly not new phenomena, their impacts on birds may be worsening. Changes in the earth's climate have led to less predictable weather and more intense storms, and there is mounting evidence that such incidents have contributed to recent population declines. Birds' evolutionary wisdom may no longer be keeping pace with the times.

By morning, the winds have eased, the sun peeks out from behind a bank of clouds, and we're anxious to get on our way. We scan the horizon for whitecaps. We listen to the marine forecast on the radio. Because we're barely across the border into Alaska, we pick up two different weather bands. The Alaskan version states a small-craft advisory with moderate winds and four-foot seas over a broad coastal area south of Ketchikan. Not ideal, but we've rowed in similar conditions on many other days. The Canadian forecast, always more specific and perhaps a tad overly conservative, threatens another gale that could bring fifty-mile-per-hour winds and seven-foot seas. In nontechnical terms, a gale typically means big, scary water, while a small-craft advisory leaves us on guard but not necessarily on land.

For the past five weeks, we have lived by some version of the Beaufort scale. We crouch around the weather radio listening to the forecast like it is our divining rod. We wait for the crackling voice that will read our daily fortune. Do we stay or go? Can we row safely? For how long? In northern waters, marine forecasts often cover such a large area that any advisories require a heavy dose of local interpretation. So each day we listen like it matters, look out at the ocean in front of us, and make the call. Decisions are easy when they are definitive. But the in-between is harder. Today we sit squarely on the fence. We listen to the forecasts again and opt for the more optimistic Alaskan version. We *are* in Alaska, after all.

After packing up camp, we launch our boats into the calm channel between our mainland campsite and an island just offshore. I'm suddenly exuberant in the sunshine, reveling in the way my boat slices through the clear water. The muscle fibers in my limbs begin to lengthen after too many cramped hours in the tent. I hear the sharp exhalations of porpoises before I see them, framed by the deep blue of the open Pacific. As I watch their playful forms arc through the water, we meet three-foot rollers around the first bend. Though the waves aren't fierce or even particularly big, they startle me nonetheless. The sea surface appeared smooth through the binoculars. This discrepancy reminds me that I am still guilty of seeing what I want to see.

Pat and I ride the waves side by side in silence. This has become our habit lately when we meet unfavorable conditions, neither of us daring to give voice to our anxieties, hoping that if we ignore the obvious, it will go away. I begin to examine the shore more closely. Before leaving, we'd pored over the chart. It showed a series of irregularities along this ten-mile stretch of exposed coastline, but we couldn't tell if there were any beaches where we could land our boats. So far, the prospects don't look good. Small islets guard the entrance to any features that might offer protection. Swell crashes into "boomers," submerged rocks that explode with the surf. They threaten like mythical dragons of the sea, sucking huge, hungry gulps of water before roaring in a boiling rage of foam and froth. With the limited vantage of rowing backward, we risk not seeing a boomer until it is too late, rolling our boats and smashing us against the reef. The more I look around, the more I realize that this would be a spectacularly bad place to capsize.

For centuries, embarking on a sailing voyage involved lengthy negotiations with the whims of the sea. At a time when the world might still have been flat, an elaborate set of mariners' rules offered sailors the hopeful promise of safe passage. No bananas on board. Don't leave on a Friday. Avoid redheads before setting sail. Never change a boat's name. Don't whistle in the wheelhouse. It was true that those sailors unlucky enough to be stuck with a cargo of bananas often never returned—in the 1700s, most ships that sank carried with them a load of the Caribbean fruit. Deadly spiders occasionally emerged from bananas, striking a man dead in a moment. And the temperature-sensitive fruit can ferment in the heat, releasing toxic levels of methane that routinely poisoned anyone trapped in the hold, including imprisoned slaves. Many of the other rules had little to do with facts. In the wrath of a hurricane, the boat is the only character that matters, and like any character, it deserves the respect of a proper name. Christ was crucified on a Friday; no need for irreverence when a crew's lives are at stake. Whistling promises to stir up the sea's wrath. Redheads do the same. Even today, with the benefit of detailed satellite imagery and sophisticated weather models, modern mariners steer by equal parts superstition and science. I am no exception, though today the superstitions speak more loudly.

Rowing through the building waves, I try to imagine which of the seafaring rules we might have violated. No fruit, no redheads, it's a Monday, my lips are too taut to whistle even if I wanted to. Instead, I recall a statistic about avalanches and other mountain hazards—groups traveling with women are much less likely to have accidents that result in fatalities. Whether due to female intuition or simple caution, the outcome is clear. I doubt the same statistics exist for rowing, but the message won't leave me alone.

"Pat," I yell across the waves. "We need to go back. Something doesn't feel right to me."

"Are you sure?" he asks. "It doesn't seem that bad right now." He looks past me at the ocean ahead. He has never been one for superstition.

"It's bad enough," I shoot back with a confidence I don't feel, "and we'll be screwed if we get stuck out here. If it calms down, we can try again later." There's no hope of rescue along this remote stretch of coastline. In fact, it's unlikely our handheld VHF would broadcast far enough for anyone to hear. We haven't seen another boat for days, and it's too early for the fishing fleet to head north. Using the satellite phone, which isn't waterproof and is often slow to get a signal, would be nearly impossible in these waves. We're completely on our own. I want to go back to shore, now. It's clear that Pat wants to continue.

This isn't the first time our danger radars have crossed signals. Rarely does Pat regard a situation as dire. Never have I heard him say he might have died. I'd like to think I'm not prone to irrational panic, but I can't help but chase fear to its terminal conclusion. Today, I see waves in the cockpit, rowboats upside down, two tiny bodies in a brutal sea. I see my dad's best friend thrown from his boat, sucked down a river that once seemed friendly. I see my mom's terrified expression as she stands on the bank of another river the next time we go fishing together. I see satellite tracks vanish from the screen when a bird meets its end in a storm. I see enough to know that we need to go back.

Pat says nothing in reply but follows my lead and turns his boat around. Within a few minutes, the wind begins to tug hard on my braids and I feel the first real twinges of worry. The contours of the shoreline bob and lurch erratically with the rowboat's motion as spray covers my arms and face. Mantra-like, I recite a litany of reminders about seamanship, rowing, cold-water survival. Don't let the boat get broadside to the crest. Stay loose, keep the blades in the water. I cling to each whispered syllable as desperately as I grasp the oars.

Glancing up, my eyes focus on Pat's boat close behind me. After only a few seconds I force myself to look away from the violent yawing of the sleek, blue capsule. Nearing our old campsite, we row hard to gain ground as the tidal currents pull us away from land. Waves break near shore, prompted to curl in the shallows. Finally, my boat crashes to the beach and I jump out of the cockpit into the turbulent surf. My feet welcome the uneven surfaces of schist and sand. I pull the boat up as far as the water's buoyancy allows and begin to unload.

"Turning around was a smart call," Pat offers in apology as he lands next to me, hair tousled and wet. I tell him I'm just glad to be out of the water. I don't mention that I'm also relieved my decision was the right one, that I didn't end our day on cowardice alone.

Back on the safety of land, we sit and watch the weather. By late afternoon, the full force of the promised gale arrives. The Canadians were right. Soon the ocean buoy reports are of thirty-foot seas and hurricane-force gusts. As the wind begins to howl again, Pat walks to the end of the point to photograph the waves. I retreat to the tent and pull out another book, this time a novel. We don't bother listening to the updated forecast—the sea has told us all we need to know for now.

As we wait out the storm, which builds in fury each day, I'm amazed at my body's capacity to be sedentary for hours on end. Over the course of nearly eight hours, I realize that I have climbed out of the tent only three times—to pee, to heat water, and to adjust the tent's guylines. From constant motion to a slight step above comatose.

Finally, late on the third night, the sound of the flapping tent is replaced by a chorus of geese. I lie awake in the stillness as hundreds of birds pass overhead, their voices multiplying as the wind eases and then stops completely. The next morning dawns as though it has always been calm and sunny. Even the more conservative Canadian forecast gives us the go-ahead to leave. We pack our boats lazily on the fine sand; gone is the urgency brought by pounding surf and breaking waves.

As we near Ketchikan a day and a half later, the sun is still shining. In a place that receives 150 inches of rain a year, a warm Sunday afternoon is cause for celebration. The air temperature has peaked at fifty-one degrees Fahrenheit, rising just six degrees above that of the water, and the effect is astonishing. Kids in bathing suits line the shore. Two girls dog paddle in the shallows, dunking their heads, ponytails slick and dripping. A man in boxer shorts floats past in an inner tube. The enthusiasm is contagious, and we bare more flesh than we have in months. No longer hidden by layers of clothing, I notice how my body has changed since we left Bellingham six weeks ago. I am all wire and rope, tendon and muscle. My legs are hairy and bruised. My arms are tanned only at the wrists, in the space between my shirtsleeves and my gloves. Pat, already brawny by nature, has biceps that look cartoonish.

When we arrive at the public dock, even the customs agent is smiling as she asks us a few distracted questions. A fisherman plays a peppy tune on his saxophone, relaxing on the scuffed deck of his blue and white boat. Rain is in the forecast, but at this point no one cares. We pull our boats onto the dock and wander through town, swapping grins with the locals. We are here before the tourist season has begun, and rates are cheap at a hotel that overlooks the marina. So we decide to scrimp on a meal and splurge on a bed. We cook pasta over our camp stove in the room, and check e-mail with the free Internet. An article I submitted to a scientific journal before we left has been accepted. It feels like a message from another life, and I'm less excited about this news than the promise of ice cream from the nearby grocery store.

The next morning, we pack our dry bags again, donning dirty clothes over clean bodies, and launch from the harbor under curious stares from a fisherman mending his nets. We have forty miles of rowing ahead of us, but it turns out to be a relatively easy day, with light winds and few waves.

By late evening, we stop to camp at a beach covered with crushed shells. In the distance is the silhouette of Prince of Wales Island, suspended in the long northern dusk. As I light the stove to heat water for tea, I hear muffled groans coming from the water's edge. Over the previous several hours we had spotted several bears along the shoreline and I tense at the unfamiliar sound. I call to Pat as he sets up our tent in the forest. He walks down the beach to join me and a moment later there is an obvious slap on the water in front of us. Ripples undulate in concentric rings. Another splash, a plume of water followed by spray barely visible in the twilight. The rising tide laps inches from my boots. As I watch the water's surface intently, bubbles begin to rise. Suddenly, a ridged and dripping mass of four heads breaks the surface, mouths parted.

"Pat, I think they're bubble-netting!" To feed, small groups of humpback whales sometimes cooperate by swimming head-to-tail, forming, as the name implies, a giant net of bubbles. This is a piece of magic I've read about but never seen in person. With controlled breaths, whales etch circles in the water, enveloping schools of small fish—herring, salmon fry, or candlefish—before lunging to the surface and opening their mouths to swallow thousands of gallons of seawater teeming with fish. We stand watching until cold water pools against our feet, and darkness clouds our vision.

After dinner, we relax around the campfire, listening to the whales. As we poke the embers with driftwood, streaks of electric green begin to skitter across the sky, a late-spring display of the northern lights. Soon, night will disappear entirely, replaced by summer's midnight sun.

Later, I lie awake in the tent, listening to Pat's steady snores, echoed by the rhythmic lapping of the tide. Tonight we have returned to our first date at Pat's cabin, to those long dreamy days in *Sprucey*. Tonight I don't need to know the answers—to whether or not we should have a baby, or what it means to be a biologist who has strayed from wonder, or even whether we will make it to the Arctic. I need only to be here, breathing in time with the waves.

PART THREE

Yukon

INTO THE
MOUNTAINS

After nearly twelve hundred miles of rowing, we arrive at a destination we have been craving since we left Bellingham: the log cabin on Lynn Canal that Pat and I built together. The cabin that built *us* as much as we built it.

This season, our cabin is only a temporary stopover where we will leave our rowboats and resupply for the next leg of our journey: a crossing of the snow-covered Coast Mountains. Soon, we must gather the courage to load packrafts with skis and mountaineering equipment and launch across eight miles of ocean in boats much less seaworthy than those we've been rowing. But, for now, we focus only on the luxuries in front of us: soft bed, hot fire, real food.

After weeks of being tossed around in the springtime surf, Pat and I nestle indoors and happily ignore the weather. At night, we relax against the cotton sheets as branches rustle and raindrops spatter on our metal roof. During the day, we scowl at sun rays peering through the fog and shrug at frothy waves crashing on the beach. We park ourselves next to the woodstove and watch the clouds zoom past, staring at the row of jagged, glaciated peaks that line the canal.

One lazy afternoon, as I scroll through e-mails I downloaded at our last town stop in Juneau, I notice a message from the National Science Foundation. It can be only one thing: the decision on my fellowship application. The funding for these awards is competitive, and before I click on the message, I dismiss it as a form rejection letter. But as I scan past the salutation, I see the words *Congratulations, you have been selected to receive a post-doctoral fellowship.* Despite the unexpected good news, I don't feel any of the same excitement I had when I first learned I'd received funding for my Ph.D. research. Besides a flash of pride that I have convinced a panel of judges I had a decent idea, I don't feel much of anything, except a little panicked at the thought of having to make a decision. Academic successes seem the stuff of another world.

The afternoon before we leave, we take a walk in the woods behind our cabin. Just as with the first time we saw this forest, I'm smitten by the muted colors and tree-cast shadows, light filtering through the canopy. We follow the local bear stomp—a trail of footprints worn into the moss by generations of grizzlies stepping neatly in one another's tracks—and pass a stand of cottonwood trees, where I stop to scan for eagle nests. As I'm looking, I notice a small bird hammering against the bark of an adjacent tree. I focus my binoculars on the bird for a closer look. What I see surprises me. A slate-colored back. A white patch on the cheek. A striking black cap.

Black-capped chickadee. It's the species I've spent the last several years studying. It's also the first one I've ever seen near our cabin. Chestnut-backed chickadees are common in coastal woods, but a black-cap, partial to birch forests farther inland, is a rarity here. I feel an odd mixture of elation and regret at the sight of this familiar little bird. It's like bumping into an estranged childhood friend, a reminder of a fond relationship that went sour. I instinctively examine the bird's beak for signs of a deformity. I'm relieved to see it looks perfectly normal. Then my mind flashes to the steel cages and lifeless bodies of the birds I dissected, and I feel again the deep ambivalence about my research, the uncertainty about the pending fellowship offer.

But as the chickadee comes closer and begins to scold me, I can't help but laugh. Its message is clear: don't mess with my forest, or my kind. No matter where you encounter them, these small birds are all vim and vigor, as sure of their place in the world as any self-important humans. I stop fretting about research and jobs and simply watch. The bird dangles upside down as it searches for insects, then flits to a branch just overhead. It begins again: *Chick-a-dee-dee. Chick-a-dee-dee. Dee-dee.* Chickadees' calls vary in tone and duration. When responding to a predator or an intruder, the number of *dees* reflects the perceived level of threat. Today, I only warrant two *dees*. Hardly more than an annoyance. The fact that this bird cares so little about me is reassuring. I don't know what these next months will bring—whether our time in the wilderness will carry me back to science, or if acceptance of a prestigious fellowship will be enough to curb my wanderlust. But one thing is certain: I can always count on the brashness of a chickadee to keep me humble.

On the morning of our departure, clear skies pry us from bed at 5 a.m. to begin our journey again. I moan as my watch alarm sounds its irritating bleep. Once we leave, gone will be the warm glow of the woodstove, the endless coffee, the perfect view, the crab that we shelled for dinner last night. In their place will be a cramped tent, thin sleeping pads, and mushy instant oatmeal. As much as we want this trip—have poured every ounce of our beings into it, in fact—we're also completely human. Bed or bedroll? Couch or snowfield? Safe cabin or mountain pass? The real question quickly becomes, *What were we thinking?*

But after five days of cabin comforts, we've run out of excuses to stay. The six feet of snow that fell recently has had time to settle. The rain has stopped and the forecast promises better weather on the way. Our plan is to leave our rowboats at the cabin and cross Lynn Canal in packrafts—small, inflatable dinghies that can be rolled up and carried on our backs. Their portability will allow us to paddle across the ocean and then continue into the mountains. On the other side of the canal, we'll collapse our paddles, strap our rafts onto our packs, and begin to hike uphill. Once we reach the snow and ice of the coastal glaciers, we'll ski across an unpatrolled border into Canada's interior, eventually reaching the headwaters of the Yukon River.

We launch from our beach at low tide, navigating slippery kelp in ski boots as we haul our backpacks and rafts to the water's edge. The wind is calm when we leave, but as soon as we duck out from behind the protection offered by the curving shoreline we face a stiff north breeze. Burdened by the heavy packs and skis lashed to their bows, the rafts are sluggish in the waves. The benefits of these lightweight boats are quickly overshadowed when the wind begins to blow. With inflatable pontoons only slightly more streamlined than an inner tube, even a slight headwind leaves us barely treading water.

We pause to debate our options, running through a handful of different scenarios and all of their potential outcomes. "Maybe these are just local gusts," I say hopefully. "It's not supposed to be windy today."

"We could head toward the islands—they might offer some

protection," Pat replies. "But we don't want to be stuck out there."

"Maybe we should hug the shore as long as we can. It's not the most direct but at least we could land if we had to."

"There's always the option of waiting a few hours."

"What if it just gets worse?"

This is the iterative manner in which Pat and I usually make decisions, talking through alternatives, stating the advantages and disadvantages of each, and then repeating the process until a pattern emerges—or until we get sick of waffling and flip a mental coin. But only rarely does one of us offer a hard-and-fast answer. Only rarely does a hard-and-fast answer exist. If we waited for perfect conditions, we'd still be sitting on a beach somewhere in British Columbia.

As we drift backward in the wind, mulling over the decision of whether to continue or retreat, a whale breaks our impasse. In the distance we see an impressive splash, followed by two more. A humpback launches its thick gray body almost completely out of the water each time, falling back on its side with a reverberating slap that echoes across the canal. We watch for several minutes as the whale breaches and rolls and wags its broad pectoral fin. No one knows exactly why whales breach. Marine biologists have proposed a variety of possible reasons: to remove parasites that attach themselves to a whale's skin, to communicate over long distances with other whales, to stun schools of small fish. Today, as I observe this forty-ton animal burst from the sea at nearly twenty miles per hour, gravity be damned, I have a different idea. It looks, quite simply, like fun.

Like us, animals love to play. Dall's porpoises will bodysurf the waves created by a passing ship's wake. It's not only humans who thrill in zooming downhill on skis or a sled; ravens and otters do the same, careening again and again down a slippery slope. Bowhead whales roll logs along their bellies like oversize toys, and bottlenose dolphins toss their incapacitated prey from snout to fluke in a solo game of catch. I can think of no reason why humpbacks wouldn't be similarly inclined to goof around.

While I'm watching the whale, I notice that the water appears much calmer ahead. I pass my binoculars to Pat for a second opinion. "I don't see any whitecaps out there, either," he says. "I think it's worth a shot." Fifteen minutes later we near the point where we saw the whale surface. The wind has eased. All signs of cetaceans have also vanished, but the tingle of knowing that an animal six hundred times my size could explode from the surface has not. Finally confident in our decision to continue, I allow my fingers to relax their desperate grip and begin to enjoy the morning paddle.

Soon, a rowdy group of teenage sea lions joins our crossing. They toss their sleek bodies through the air and dive beneath our boats in the clear water. As I watch them pass just feet below us, I recall a conversation with a friend and fellow biologist several years earlier. As part of a research project studying Steller's sea lions, his job was to capture them underwater with a handheld noose. He described the usual routine: don a black wet suit very similar in appearance to the seals that sea lions sometimes eat, suit up with scuba gear, and dive in. "If it's going well, they're mouthing you everywhere. They won't bite; they just want to play," he told me.

As I count a dozen animals splashing around us, the colorful tubes of our packrafts suddenly feel more like beach balls and less like boats. Even if they are only interested in playing, a nip with the sharp teeth of these seven-hundred-pound carnivores could easily pop our rafts. Between their acrobatic bouts, I keep a watchful eye toward the water. I have no interest in swimming with sea lions today.

Another two hours of paddling brings us to a steep shore-line on the far side of the canal. The rocks glisten, slick with seaweed. As I step out, blue mussels and small white barnacles crunch underfoot. Pat helps to stabilize my boat as I lift my pack over the edge of the rocky shelf. When it's his turn, I return the favor. We roll up our rafts, repack our bags, and grimace as we hoist our packs onto our shoulders. Our bodies are strong from rowing but spoiled by the fact that, until now, the water's buoyancy has carried much of our load. Today I stagger under the sixty-five-pound heft of food, skis, crampons, rope, harness, packraft, and paddle. Before walking toward the forest's edge, I offer a silent goodbye to the gray-green waters of the Pacific. After almost two months in the constant company of the sea, we will head inland, traveling north into Canada's Yukon. If our mountain crossing goes well, we'll be in Whitehorse in less than two weeks.

To reach the snowline, several thousand feet above us, we must first climb up a steep forested slope. Grunting, we leave the coast to crash through a band of brush and begin to make our way uphill over the mossy ground, leaning heavily on our poles with each step. In mid-May the forest is just beginning to emerge from winter. Young shoots of devil's club sprout from the recently thawed ground, and yellow-green lichen dangles from tree branches like neon tinsel. This is my favorite season in Alaska, when the land swells with the promise of all that is fresh and new.

Shirts drenched in sweat, skis on our backs growing heav-ier with each step, we stop for water and snacks. I finish my granola bar and lean back against the soft duff, stare up at the dense canopy, and listen. All around us, forest birds are welcoming spring. Perched on a branch above our heads, a Pacific wren launches into an enthusiastic outburst. These birds have lungs the size of lima beans but their voices are as large as their organs are small, filling the thick air with a waterfall of sound. "I hear a varied thrush," Pat announces proudly as a long, monotone whistle erupts from a stand of shrubs nearby. I quiz him on the other bird songs I've taught him, memories of which easily become fuzzy after Alaska's long season of winter silence. A chorus of Townsend's warblers exchanges buzzy notes and wheedles. A golden-crowned kinglet whispers its quiet *tsee-tsee-tsee*. A hermit thrush calls from the understory as a brown creeper scurries up and down the trunk of a nearby hemlock.

I have studied some aspect of the ecology of each of the species we see. I can tell you where these birds have spent their winters, how many eggs each will lay, and how they raise their young. I know that a Pacific wren depends on old-growth spruce and hemlock trees and feeds on insects hatched from salmon streams. I've counted the first Townsend's warblers as they arrived from wintering areas in Mexico or Central America. I've found the nests of varied thrushes, blue eggs cupped in a bed of grass and moss. I've also stared at a computer screen for days trying to determine why a warbler likes one patch of forest better than another; I've painstakingly counted seeds and analyzed blood samples in the lab in an effort to answer the question of what a bird had for dinner. Tedium is a regular part of my job. But today the graphs and calculations fall away as I inhale the scent of dirt and spruce needles. Out here, I am half scientist, half disciple. I've left the laboratory far behind and, with it, the need to quantify and contain. In its place, I've reconnected with the simple act of observation.

Watch. Listen. Learn. These were the tenets of the earliest naturalists, the instincts of indigenous people around the globe whose survival depended on knowledge gained from the land. My passion for birds and the natural world also emerged from these basic principles. I fell in love with the image of the sky blocked by kittiwakes flying in perfect synchrony, became intrigued by a chickadee whose beak promised to foretell something about our environment, tuned my ears to the nuances of a warbler's song. What drew me to research was not the rigor of statistics or the mystery of what lies within an organism's genetic code. It was the birds themselves.

Observation can guide us to wonder. It's also the foundation of all scientific inquiry. Without observation, we have little hope of understanding an individual, a species, or an entire ecosystem. But for most contemporary scientists, including myself, this is no longer enough. Innovative technologies have made classical scientific techniques—many of which relied on an observer's eyes—obsolete. It's not necessary to spend hundreds of hours peering through binoculars to determine what a bird eats; a genetic test of its feces provides the same information. Aerial imagery offers real-time views of the landscape, eliminating the need to survey every acre of every meadow. Satellites relay data about animal movements to our desktops from thousands of miles away. Even social media and Internet platforms have changed how we do our work, adding millions of observers around the globe and providing sample sizes much larger than anything a single study could achieve.

Science has gone the way of most other things in our digital world. High-tech, computer-centric, and data-hungry. As a result, we know much more than we used to. But we also spend much less time as observers. Wandering through the woods with only a backpack, a notebook, and a pair of binoculars has become a novelty, rather than a necessity, for many biologists. After five years of laboratory research, it's what I need most right now.

Among the warblers and wrens, I force myself to think about what the National Science Foundation fellowship offer might mean. By most measures, it's an exciting opportunity. It would be a chance to launch my career as an emerging researcher. I would be surrounded by scientists making discoveries that shape our collective knowledge and have resources not available in Alaska. There's only one problem. I'm not yet sure what an Ivy League school might offer in the way of wonder.

As we gain elevation and wind through a quilted landscape of snow and moss, we alternate between skiing and walking. Skis go on our backs, then on our feet, then on our backs again. We shimmy over logs and under alders, contorting our bodies to meet the organic shape of the forest. The higher we climb, the more spring fades back into winter. Birdsong quiets. A continuous blanket of snow covers the ground. By late evening we've made it to the toe of the glacier that connects to the ice field leading into Canada.

The pass we studied from the cabin looms large and intimidating above us. Massive chunks of snow and ice—evidence of recent avalanches—lie jumbled beneath the adjacent slopes. We had seen this debris with our spotting scope, but up close the slope angle looks even steeper, the avalanches scarier and more destructive. It's been several days since any snow fell, and we're gambling on the assumption that there has been enough recent warming to flush the rotten snow off of slopes that might be prone to slide. Right now this feels like a big assumption.

Pat and I are both tense, and we go through the familiar routine of setting up camp without speaking. Pat erects the tent, then pulls sleeping bags and pads from our backpacks. I light the stove and begin to melt snow in our cookpot. When we have each finished our jobs, we crouch on our backpacks and devour pasta slathered with butter and cheese. I imagine us in our cabin, sitting at the table with driftwood legs, passing the salt shaker and drinking wine from a box. I'm trying not to think about the fact that we won't enjoy such luxuries again for several months. But mostly I'm trying to ignore the intimidating pass we will have to cross tomorrow. When I ask Pat what he thinks about avalanche hazards, he tells me the conditions don't seem as solid as he'd hoped. I know exactly what he means, but hearing him say it aloud makes the macaroni taste bitter in my mouth.

The mountains operate on a sliding danger scale. There are no right answers, only better or worse choices. Here, we have no guidebook to follow, no avalanche forecast, no one to ask for route information. When we planned this portion of our trip, we queried old journal reports, searched the Internet, and probed locals for details but could find no record of a similar traverse ever having been done. I wake several times throughout the night, startled by the gentle flapping of our tent fly or the wind sweeping down the valley. In my dreams, I feel the frozen ground disappearing beneath me, then a raw, icy sting as I claw at the snow with bare hands.

The next morning, I fight the urge to burrow deeper into my sleeping bag and instead crawl out of the tent to heat water for breakfast. We inspect the slopes more carefully; no new avalanches have come down overnight and we decide to climb high enough to assess the snowpack. The pot steams in the misty morning as I open an extra packet of instant coffee. We'll need the caffeine to get us through the day.

I offer to break trail to the pass. Slowly, I pick my way up the slope, prodding the snow with my ski poles, assessing its density and structure. As I probe I feel for the suspect layers that could signal the threat of an avalanche waiting to happen—a hard crust, such as that formed by freezing rain, slippery beneath a heavy, wet blanket of snow. I switchback again and again, gaining confidence as I go. The snowpack doesn't change much as we climb, and the avalanche of my dream fades. Pat follows several hundred feet behind, giving me space to assess the snow and choose a route. He flashes a thumbs-up when I look back at him. By the time we finally reach the pass, the morning's drizzle has turned to flurries and the visibility quickly disintegrates. Like many other summits and high points, this is no place to linger. We look briefly at the map and decide on a plan. We'll descend a long, continuous slope to the glacial valley below.

In avalanche terrain, we always ski one at a time to avoid the possibility of accidentally triggering a slide onto the person below. We also serve as each other's eyes, keeping careful watch in case of an accident. If a person is caught in an avalanche, determining the general vicinity of where the victim is buried can mean the difference between life and death. Even with avalanche transceivers that send a signal to would-be rescuers, it can take minutes or hours to locate a victim, and every second a person is buried counts. If you're lucky, you'll find yourself with an air pocket hollowed out of the cement-like snow. If you're not, the cold presses in against your face and suffocation comes more quickly.

Here, because of the poor visibility and lack of any protected mid-slope viewing points, Pat and I will have to ski out of each other's sight. I hate this fact, but I also know it's the safest choice we have. Pat volunteers to go first and I don't protest. As he begins to descend, I peer into the grayness, following his red coat with my eyes as long as I can. I try hard to control my panic as I watch him easily dodge a slow-moving runnel of wet snow before he is swallowed by the fog. When I guess he has made it off of the slope, I start down, acutely aware of the fact that Pat won't be able to see me if anything happens. As I descend, I channel all of my focus into skiing. My pack is heavy and awkward and threatens to throw me off balance. Each turn hammers my already tired quadriceps. I squint through my goggles into a featureless patch of white. Partway down, I kick off a small avalanche. It stops before it musters any speed, but I'm afraid each time I cut the slope that a bigger slide will follow. When I can finally make out Pat's form on the expansive flat glacier below me, my heart lifts.

Once I reach him, I drop my pack next to his and sit down to catch my breath.

"I don't like this—" I say.

"And going higher isn't going to help," Pat finishes my thought. If we continue on our original route, we'll have to gain elevation, and avalanche hazards will likely worsen. The west-facing slope we climbed to reach the pass seemed relatively safe; the one we just skied down clearly was not. It doesn't take more than a few words to realize that we've arrived at the same conclusion. It's time for a change of plans.

Not knowing what conditions we might find, we had outlined an alternative route on the map before leaving. It will take us over lower-elevation, lower-angle terrain and force us to exit off of one glacier and climb back onto another. Typically, transition zones between glaciers and rock are places to avoid—a glacier often terminates with fins of steep, blue ice, offering no way up and no way down. The other route is also longer, which means more days of travel and correspondingly smaller food rations. But, given the unstable snow, it seems like our only option.

As we descend, we weave between crevasses until the ice

ends abruptly at a narrow, scree-filled valley. Here we take off our skis, strap them to our packs, and begin hiking again. The footing is rough and I struggle with my heavy load. The comforts of the cabin are a lifetime away as I wonder whether we will ever make it across the mountains. "What's wrong with us?" I mutter to Pat. "We spend years building a home on the coast and we can't seem to sit still long enough to enjoy it."

Several minutes later, a pebble whizzes past my head. I look up to see a mountain goat high above us as he steps onto the precipitous edge of a waterfall. Pat is a few paces ahead of me and I call to him to pay attention. The goat pauses for a moment on a tiny, vertical patch of moss and bends his head for a drink. His horns parallel the ground as he sips from a rivulet that plummets several hundred feet to the glacier below. Standing there calmly, the goat drinks again and again from the cold water.

The goat is all grace. The mountains are his home. We're only visitors here, clumsily finding our way. When the goat begins to move again, we continue downhill. Suddenly, my steps feel surer, my load lighter. I walk carefully with my crampons along a narrow tongue of ice that leads to the valley bottom. This is a place of steep slopes, avalanches, and boulders the size of cars. It is also a place of magic. As I walk, I whisper a quiet thank-you to the goat.

BORDER CROSSING

Several days later, we're camped on the lateral moraine of an unnamed glacier near the Canadian border. I wake up and peer out of the tent into a sea of flat light. I blink through gauze. I look for the peaks that were there yesterday. No mountains. No sky. Just a few boulders balanced at erratic angles on top of blue ice. Our camp straddles a ridge of crushed rock that is the only discernible feature in a field of white. We've made it back onto the ice route that will take us to the Yukon's interior, but the weather makes me feel anything but celebratory. It's snowing big sticky flakes that turn to water on my cheeks as I unzip the tent to look outside. "Nooo..." I whine as I snuggle closer to Pat, duck my head back into the nest of down, and draw the last of the residual warmth from my sleeping bag.

"Maybe it looks worse than it is," Pat offers with an annoying cheerfulness. He uses my retreat as an opportunity to get dressed, pulling on his bibs and jacket, grunting as he forces his feet into frozen boots. Our tiny, transient home offers only enough room for one of us to dress at a time. Wiggling into socks and pants, zipping a jacket, or spinning around to face the tent door requires all of the extra interior space we have.

As we clamber over each other, it's hard to avoid taking an occasional elbow in the back or jabbing an errant knee in the other's thigh. The first one out of the tent gets the easy job—lighting the stove and preparing breakfast. The other pays the price of dawdling and has to deal with packing up wet sleeping bags and a frosty tent, a task that turns hands to stiffened claws before the day has even begun.

Eyes and joints still hazy with sleep, I gather my clothes and squirm into a cold sports bra, then pull on damp socks and an extra layer of long underwear. The water is nearly boiling by the time I finish dressing and drag myself out of the tent. My knees ache as I kneel on the frozen ground and begin to pull tent stakes.

"Here, you take over," Pat insists. I muster a few half-hearted words of protest before gratefully accepting his seat on a boulder perched over the steaming stove.

"Thank you." I smile as he shakes frost off of the tent fly. Out here, generosity comes not in the form of material things, but the gentle lessening of each other's daily burden.

We gulp down oatmeal and coffee, strap on our skis, hoist our packs, and start shuffling. Wind-driven squalls blow snow downhill, and we squint through our sunglasses, the snow's reflection blinding even without a hint of sun. When Pat's in the lead, I focus on his form, stare down at his tracks, and occasionally close my eyes. When we swap places, I have to remind myself to study the pattern of my skis or glance over my shoulder to orient myself in the featureless landscape. Vertigo makes me stumble as my internal compass swings wildly.

For the rest of the morning, the sky continues to hang low and heavy. Traveling uphill toward the pass that divides the watersheds of the Pacific Ocean and the Bering Sea, we ski into a headwind that whisks away the sweat on my forehead and the moisture on my lips. The mountains are still hidden in a shroud of fog. So far there have been none of the overexposed colors that define a glacier in the sunshine, no piercing blue sky or sharp white peaks. Instead, it is flat, muted gray everywhere I turn.

With the steady *swish-swish-swish* of my skis comes a stream of loose, meandering thoughts. I picture my sister at home with her son. I wonder how my dad is really doing behind the upbeat words he offers each time I call. I think about the chickadee at our cabin, prospecting for a mate in a forest where few other members of its species live. I recall the shiny-coated black bear that passed our camp two nights ago, following the glacier to the coast.

Eventually, my mind travels far from here, to questions about what will come after this trip. I have a looming decision to make. I must respond to the National Science Foundation fellowship offer within two weeks. In the days since we left the cabin, I've wavered between excitement and dread as I've tried to imagine myself showing up on the East Coast in the fall with a commitment to develop models of bird beaks. My work there would involve building mathematical equations extrapolated from photographs of chickadees taken on the other side of the continent. What had once seemed so necessary now feels petty. I'm no longer sure what my proposed research has to do with the birds I've watched soaring past, their purpose so sure. What would it mean for Pat and me to move, even if only temporarily, thousands of miles from the place we love most? Could this journey possibly carry me back to science? When I think of committing myself, once again, to years in a laboratory or in front of a computer, distant from the birds, I feel my jaw tense and my shoulders droop. My body offers an unequivocal answer.

Gradually the sky opens just enough to reveal a glacial lake shimmering in the mist to our left. Ice shines through the clear water, and a rock is perched near its edge. As I get closer, the rock turns its head and is no longer the color of stone. What was gray from a distance has become white. I blink and reach for my binoculars. Through glass, I see a shocking scene. At nearly five thousand feet on a broad ice field, a trumpeter swan is taking a bath. A world away from the lowland ponds and lakes where it breeds, far from the plants it must eat to power its flight, the bird appears perfectly at home. It looks over at us, stretches its long neck toward the sky, then continues its ablutions. As I look down to make a note about our swan sighting on the map, I see that we are nearly straddling the border between the United States and Canada—a boundary that means little to us or the swan.

The surprise sighting reminds me to look up more frequently, and soon I realize that we have joined a procession of other travelers. First I notice a small flock of swallows as they fight against the fierce headwind. The birds toggle back and forth to switchback into the gusts. When the clouds lift above the mountaintops, I see a hawk catch a thermal and rise into the gray sky. Soon, other migrants pass by. Shorebirds call overhead. A flock of geese flies high above us. Suddenly this place feels much less deserted. It occurs to me then that our route is a logical flyway for birds heading inland from the Pacific coast. It's no accident that we're all following this path of least resistance, a low point in an otherwise jagged range, a toothy gap in the barricade of the Coast Mountains. From Lynn Canal, birds funnel up the Katzehin River and over the mountains just as we have.

On every glacier crossing we've made there is evidence of these trials. A lone feather hollowed into the snow. The skull of a surf scoter perched upon a rocky moraine. A tiny warbler resting motionless on the ice. I make guesses about the circumstances of their mishaps; perhaps a blizzard as they rounded over a steep mountain pass or a rogue storm that blew them off course. Or perhaps they ran out of fuel, the thousands of miles separating Alaska from Central America or the South Pacific eventually catching up with them. Today, the migrants we're seeing look very much alive and purposeful. As we watch, I marvel at the fact that they know which path to follow. How do they find their way over oceans and glaciers, across continents and mountain ranges, without the benefit of a map, compass, or GPS?

For many years, we've known that birds use the sun as a compass. At first glance, this concept seems simple. Even the most directionally challenged among us understand that the sun rises in the east and sets in the west. But in practice, like most things in nature, it requires more than a basic heuristic. Because the sun is not fixed on the horizon, birds must be able to adjust their orientation to account for time of day. Early experiments on avian solar navigation found that when a bird's internal clock was shifted (by changing patterns of light and dark), it compensated appropriately for the new time, launching in a direction that *would* have been correct if time had not been scrambled. Birds may also derive clues from shadows and hues, much as we do in the nuances of a painting or a photograph. But the sun can't explain it all. What about clouds and storms, night migrants, and twenty-four-hour Arctic light? Decades of research has taught us that birds use a diverse suite of tools to find their way, orienting by stars or wind, responding to signals from magnetic or olfactory receptors. In many cases, though, when it comes to explaining how birds navigate from one place to another, the truth is that we still simply don't know.

Here, among the ice and snow, lies the magic that is hard to capture through textbooks or journal articles or the blue screen of my computer. For individual birds, there is nothing formulaic about migration. Each season is different, every journey unique. Birds must learn to follow the fickle pulses of the jet stream, to negotiate snowy mountain passes, to skirt the ferocity of spring storms. E7, the bird whose satellite track revealed an impressive 7,396-mile nonstop flight, didn't attain fame because she did anything particularly unique for a bar-tailed godwit, but because she showed us how remarkable the usual can be. To get to their breeding grounds, birds chance everything in an ordinary act that is by any measure extraordinary.

It's at these times that I feel both most and least like a scientist. I am smitten by the very fact of the birds' existence. No matter how much we learn, migration will always be mysterious at its core. A part of me will always hold that same suspended disbelief that early naturalists did, 250 years ago, when they assumed that swallows must have hibernated by burying themselves in mud all winter. The alternative seemed too outrageous—wings beating high above oceans and glaciers, following an invisible map thousands of miles *just because*. Even today, with satellite tracks across computer screens offering indisputable proof, it's hard to fathom that our smallest winged companions are capable of such feats. These aren't the details that populate scientific papers or solicit grant funding, but they're the ones I hold in my heart. Today I want only to sit back and observe, to accept magic as magic, to swap knowledge for awe.

By late afternoon we reach the crest of the gentle pass that marks the high point between the Pacific Ocean and the interior. With a quick congratulatory hug, we tip our skis downhill into the Yukon River drainage. The descent off of the glacier is surprisingly gentle and we ski down to find a textbook example of the rain shadow effect. As warm, wet air masses from the Pacific hit the Coast Range, the mountains act as a giant sponge, wringing water from the sky and leaving most of the moisture on the west side of the mountains. The difference this creates is remarkable. Instead of towering Sitka spruce and western hemlock draped with lichen, we're suddenly surrounded by an arid pine forest. Wet has turned to dry. Blurred edges have become crisp. Even the sun makes an appearance, shimmering against frosty willow branches just beginning to burst with silky gray catkins.

The next day, working our way into the valley that drains the headwaters of the Swanson River, we begin to hopscotch from one small patch of snow to the next. In this transition zone, where spring is nudging out winter, there is no perfect way to travel—too much snow for hiking and too little for skiing. The river flows through a narrow slot canyon choked with ice, making paddling impossible. We clamber over logs and across fields of pine needles and crispy brown ferns, skis dangling from our feet like useless appendages. Sweating and straining, we cover less than a mile in two hours.

When the river's gradient finally eases, the steep canyon walls give way to gravel bars, and the packrafts on our backs taunt us with the prospect of a free ride. As enticing as the river looks here, we know nothing about what lies ahead. So we hike for another mile, continuing our agonizingly slow progress through the brush. Finally, the temptation is too great and we decide to try boating. We take turns pulling ashore to scout around each bend. The water churns and splashes and I grip my paddle nervously. But even with our frequent stops we move quickly downstream and soon the rapids end in a gentle, grass-lined channel. The river's murkiness disperses like a finger painting into the turquoise waters of Tagish Lake, where a single swan is bathing by the river's outlet. There's no way of knowing, but I laugh at the prospect that this could be the same swan we saw on the ice field, a bird fastidious about cleanliness.

* * *

I'm relieved to have reached the lake. I'm also hungry. We've been rationing the past several days, trying to make up for the fact that we didn't pack quite enough to eat. During our mountain crossing, with no places to refuel along the way, everything we needed had to be carried on our backs. Of course, food is essential. But food is heavy, too—two weeks' worth for two hungry people weighs in at seventy pounds. In the transition from boat to backpack, every extra pound taking its toll on our bodies, food suddenly became a limited resource.

Prior to the start of the trip, Pat had contacted the owners of a small wilderness lodge on Tagish Lake to ask if we might be able to buy some food when we passed. After a long phone conversation, Pat told me that although the proprietor had a heavy accent, he thought she said yes. My vote for carrying extra food in case there had been a miscommunication was vetoed as soon as we saw the size of our packs. "It's a lodge, I'm sure they have tons of food," Pat said. From the comfort of the cabin, paring back our provisions seemed perfectly reasonable.

But when we reach the lodge and meet the friendly young couple, I begin to wonder again if the idea of a resupply had gotten lost in translation. They greet us at the dock, give us a tour of the cabins where guests stay once the season begins, and point out a sandy campsite by the lake. There is no mention of food. Two hours later, the man walks down to our camp to invite us to dinner.

The meal is amazing. And the portions are tiny. I have never seen a piece of steak so petite and well mannered. It's like offering a can of premier cat food to a starving wolf. We try to quiet our growling stomachs as we chat until late in the evening about our trip and their unusual trajectory from

managers of big-city European hotels and upscale tour companies to living off the grid in the middle of the Yukon. Before
returning to our tent, I grab several handfuls of trail mix from
the food bag that we had stashed in the entry of the lodge.

"Pat, what sort of arrangements did you really make with
her?" I ask him. The one resupply I had turned over to him
seemed to have been completely botched.

"I don't know exactly. I told you, her accent is so hard to
understand. What else was I supposed to do? I'm sure they'll
give us more tomorrow."

In the morning we join the couple for coffee before awkwardly broaching the subject of food.

"Oh," the woman says. "I didn't know you needed some.
Maybe pasta would be good? And granola bars?" *Whew,* I
think to myself, *now we're getting somewhere.*

She pulls out a pound of pasta, half a dozen packets of
instant oatmeal, and eight snack-size granola bars. This will
barely feed us for a day. We're out of dinners and have been
rationing the last of our snacks. It will take at least three more
days to reach Whitehorse, the capital of the Yukon, where
grocery stores and our next box of food await us.

"That looks like a great start," I say. I try to hide my disappointment with a joke. "If there's anything else your clients
don't like, we'd happily take it off your hands."

Interpreting my comment literally, she replies, "You'd be
surprised what people enjoy when they think they're roughing
it. We don't have any problem going through food out here."

Pat and the man say nothing during our exchange, leaving
the negotiation to the women.

I phrase my next question with as much tact as I can
muster. "Do you think we could have a couple more pounds
of pasta or rice? And maybe a can of tuna? I think that might
do it. I know it's a lot to ask but we've run a bit low on food."

Really, this will translate into several hungry days, but I know we can make it if we continue to ration.

Last night the couple had told us, with no innuendos intended, about the high food prices in the north, and the complicated logistics involved in shuttling supplies by truck and boat. "You have a remote home, so you know what it's like," the man had said. And we do. But I also know how hungry I am.

I offer repeatedly to pay for the food and hospitality, but she tells me it's not necessary. We had always intended to buy the food, which clearly had also not been communicated and now makes our requests much more awkward. As I glare at Pat—*see, I told you this wasn't working out like you'd said*—she rummages through the pantry for two cans of tuna, a can of smoked sardines, and a bag of white rice. "Sorry we don't have more to share," she says, "but we will be tight on our supplies this spring."

We won't starve, but the next seventy miles of paddling will be largely consumed by thoughts of food.

The next night, a smattering of shorebirds is staged along the mudflats as we paddle by. Several whimbrels watch us from the lake edge, their stocky bodies and characteristic downcurved bills obvious even from a distance. I will learn later from a fellow biologist that whimbrels outfitted with satellite transmitters are paralleling our route from Washington. Or perhaps we're paralleling theirs. We overlap in space and time, but this is where the similarities end. These birds are tough in ways I can't begin to fathom. They started their trek thousands of miles farther south than we did. They carry nothing but fat reserves. We labor under our burden of food and gear but still feel achingly hungry. They fly into storms while we wait them out. As I watch them through binoculars gathering all the food they need from the mud, I'm equal parts inspired and envious. They make it look so easy, so natural. Tonight I'm stuck with an empty belly and a grudge toward Pat.

I'm restless in my sleeping bag, still hungry after our dinner, which consisted of a cup of white rice and a can of tuna, and unable to ignore the impending decision I must make about the fellowship. I'm finding fewer and fewer reasons to go. What good would it do to work at a highly regarded institution if I don't care about my research? Readying myself for an academic career that will take me far from wilderness feels more like a curse than an asset. Plus, when I try to picture Pat in a New Jersey suburb, I see a bird with clipped wings and no place to hide.

Just before we left our cabin, I also got notice that the federal job in Anchorage was mine if I wanted it. A government research job would come with its own strings, but at least it would be in Alaska. I know my colleagues would be cut from the same bird-obsessed, field-savvy cloth as those who led me to biology in the first place. What other form of employment would yield co-workers who spend their lives chasing shorebirds and ducks through some of the most unpleasant terrain imaginable? No one is in it for the glory or riches. Plus, for all my reservations about returning to research, I can't dismiss the fact that seeing a bird in the springtime is made more amazing by virtue of knowing just how far she's come.

Eventually, I fall asleep to the hooting whistles of whimbrels. In my dreams I see birds crossing over a sea of ice, Pat and me paddling down the Yukon River, my nephew napping in my sister's arms. For now, these truths matter in ways that academic pursuits do not. When we wake up the next morning the whimbrels are gone, leaving us to chase them on their journey to the Arctic.

YUKON DAYS

By the time we pull into Whitehorse, I've made a mental list of all the foods I'm craving—freeze-dried dinners and instant oatmeal are not on this list. Ice cream, fresh produce, and cheese are. We can't justify paying the exorbitant tourist prices for a dinner of reindeer sausage and Alaskan salmon, so we tromp to the grocery store in our ski boots instead. Wandering through the shiny aisles, we grab greedily at the shelves. Bagels, cream cheese, avocados, tomatoes, chips, cookies, chocolate, bananas. Everything looks delicious. We fill our baskets until we can't carry anything more. After checking out, we devour our sandwiches on a city bench and wash them down with several bottles of beer. The public campground is crowded, so we find an unofficial site by the park and sleep hard on full bellies.

In the morning, after a hearty breakfast of granola and yogurt, I down a quart of chocolate milk. We hit the ice-cream parlor minutes after it opens and I'm already dreaming about what's for lunch. I head to the library to check my e-mail while Pat makes another trip to the grocery store for resupply items. I start with the only message I really need to write. I click on the header from the National Science Foundation and begin to type. "Thank you for the generous offer of post-doctoral support. I am honored to have been selected. However, my situation has changed since I applied and I will have to decline the fellowship. I appreciate all you do for the sciences and for emerging scientists." After two weeks of stewing over this decision, I'm relieved to let it go. I'm still not sure if the job in Anchorage will be the right choice or if I will be ready to return to research of any kind, but I'll stay in Alaska, where I belong. I can hold on to the promise that even in the largest city in the state, wilderness exists just beyond the confines of my office door.

As I finish with e-mail, Pat shops for our next Yukon River leg like he's preparing for a wedding reception. I find him at the checkout with a forty-two-pack of white dinner rolls, four extra-large bags of chips, several boxes of off-brand chocolate chip cookies, three pounds of cheddar cheese, two heads of lettuce, and a dozen apples. He has already picked up a three-liter box of cheap wine at the liquor store and is grinning hugely as he snacks on the first box of cookies. "This is supposed to be our vacation, right?" he asks.

At other resupply points we have supplemented the boxes of dried food we shipped to ourselves with fresh items such as cheese, butter, a day or two's worth of produce, a few treats, and a celebratory six-pack of beer. Along the Inside Passage and in the metropolis of Whitehorse, splurging has been easy. But our options will begin to shrink as we head north. Produce will soon disappear from the shelves, and cheese, if any, will be limited to plastic-wrapped individual slices that cost more per pound than the finest lobster at a New York restaurant. After our next stop, at Dawson City, we will no longer be able to reward ourselves with beer—all of the Arctic communities we'll pass through have banned the sale of alcohol. So we indulge while we can.

Pat pens a quick postcard to his grandparents, now a regular part of our town ritual. He asks me to check the text he scribbled on a piece of scratch paper for spelling errors.

"We're on the famous Yukon River. We've got it easy compared to the gold rush miners. We crossed the mountains from the ocean and are heading inland now. Maybe we'll see some grizzlies. Love, Pat."

A canoe trip along the Yukon River from Whitehorse to Dawson City is one of the northern classics. In the peak of the summer season, the river hosts thousands of visitors and the best campsites go fast. The guidebook details every mile, leaving little for us to figure out. Typically, we'd be less interested in joining crowds of Boy Scouts on a prescribed route than exploring unnamed glaciers by ourselves. But after weeks of tough travel, a vacation is exactly what we need. And a canoe is the perfect way to enjoy it. Sitting on elevated seats with plenty of extra space will be luxurious compared to our recent days in overloaded packrafts. The streamlined canoe will also be much faster. In remote areas we would just accept the packrafts' shortcomings. Here, renting a canoe is relatively inexpensive and logistically simple—all we have to do is pick it up in Whitehorse and drop it off in Dawson City. After a visit to the post office to retrieve our food boxes and mail the postcard, we're ready to find our boat and hit the river.

Lined up on the riverbank next to two dozen others, the cheap white rental canoe is so battered we can see shadows of the water through the hull. Pat immediately dubs her "Casper." She's the perfect craft for a section that we previously mocked but now can't wait to begin. Before we leave, I call my parents on the pay phone near the boat launch. They're planning to meet us in Dawson City to shuttle our unneeded gear back to Anchorage and deliver our next resupply to the Tombstone Territorial Park headquarters in the northern Yukon.

As we planned our route, we faced two major logistical hurdles in the transition from glacier crossing to river boating, and then to hiking and packrafting. First, we had to find a way to transport our skis, poles, plastic ski boots, axes, harnesses, ropes, and other mountaineering equipment back to Anchorage. We also needed a resupply somewhere between Dawson City and Fort McPherson as the food required for the three weeks it will take us to travel between them is more than we can carry on our backs. When my parents agreed to meet us, and then drive the extra leg to deliver our box to the park headquarters, they solved both of these problems. And they reminded me of what selfless love looks like. This shuttle will require a thousand miles of driving, meaning that they will probably spend more time in the car than with us, but they didn't hesitate when I asked them to help several months earlier.

"We can't wait to see you," my mom says.

My dad pipes in from the other receiver. "We're all ready to go. Just let us know when to start driving."

After we launch from Whitehorse, we alternately paddle and lounge in our canoe, letting the current do our work for us as we plow through a barrelful of treats—rolls with thick slabs of cheese, chips, cookies, apples, and chocolate. The granola bars and peanut butter from our resupply boxes mostly remain untouched. We entertain ourselves by trying to recall each campsite we've used since we left Bellingham. On the particularly easy sections of the river, we take turns sunbathing and reading while the other person paddles.

With so much idle time and little need for checking maps or navigating rapids, our minds begin to wander far beyond this river, and this trip. We talk about my decision to decline the fellowship and the possibility of a research job in Anchorage. Pat mulls over a house project someone has asked him to do when he returns. Finally, we circle back to the big, lingering question. Kids. I am thirty-three, Pat is thirty-two. We still have time, but we're no longer floating through our twenties, with the prospect of a family still a decade away. Since my nephew was born, I've been thinking about the topic more than ever. Pat and I always assumed that we would have children, but now that the decision feels imminent, it's easy to come up with pages of excuses for why we should wait just a little bit longer. There are many more mountains to be climbed, dozens of places we'd still like to explore. There are the demands of a career, the practicalities of finances and field schedules. A baby doesn't fit easily with any aspect of our lives.

Besides, I've never had any visceral maternal urges. I don't hear an infant cry and feel a tug on my heartstrings. I don't see pregnant women and wish I were in their shoes. In fact, I'm terrified of losing my body to pregnancy. I've seen the transition occur in enough friends to know that where the body leads, the mind often follows. From backcountry to baby nursery. Marathons to stroller walks. Ph.D. training to sleep training.

While we were at our cabin, I called my sister every day. Each time she answered, I could hear some of the same joy that was in her voice shortly after my nephew was born. But I could also hear the exhaustion. Her current adventure is one of sleepless nights and dirty diapers, a house littered with dishes and laundry and the running shoes she isn't ready to use again. I didn't mention the fact that I felt stronger than I ever had. I didn't ask if she had any regrets. But after each conversation, I tried to imagine my own body recovering from childbirth. Or burdened by an extra thirty pounds. Or tied to an infant and then a young child. I cringed at the possibility of staying home while Pat went somewhere without me. I could think of no easy way for us to continue doing what we love, together.

In some ways, Pat seems more ready for kids than I am. "I guess we could wait another couple of years, but I don't see the point in delaying forever. Everyone says there's no perfect time," he tells me. I point out the fact that the same rules don't seem to apply equally to dads. I'd like to think Pat and I could do it differently, but there's no escaping the facts of biology. I know I can't have it all.

"There are plenty of adventures we could do as a family," Pat says.

I try to picture us on a family expedition. It would be unlike any trip we'd ever done before. What sorts of crazy logistics are required to keep kids safe and entertained on a glacier or a river? At the least, we would need a big tent, lots of food, and a way to stay warm. Even with such amenities, it's hard to imagine what we might do with a child who is too young to ski or paddle. And what of the risk? It's one thing for us to negotiate avalanches and rapids, but another to do it with children.

Clearly, there's plenty to give up. What is less clear is what we might gain. The benefits all seem hypothetical, a blind faith in other people's promises that having a child is worth the sacrifice. Even Ashley, only two months into parenthood, told me this. Hearing her report that some days it's all she can do to get showered and dressed makes me wonder at what cost these rewards might come. But it's the fear of regret that makes us return to this conversation again and again. Twenty years from now, will I be sitting on another boat somewhere wondering what I might have missed? There are certain things in life that don't offer second chances.

Of course it isn't just parenthood that might change everything. For now, our joints tolerate the abuse that comes with the terrain. We've been lucky, or careful, or both, so far escaping the casts and surgeries and crutches that so often result from accidents. For now, we are healthy and relatively young. But there's no guarantee about how long our bodies will last.

By late afternoon, the relative boredom of the river combined with too many unanswered questions makes us both irritable. We leave our conversation about having a baby behind for now.

My parents have no way of knowing exactly when and where we will arrive. Still, I almost expect to see them waiting at the bank when we pull into Dawson City several days later. As we drag our canoe up the muddy landing, I scour the boardwalk for an athletic-looking couple in their early sixties who are likely to be chatting with local kids or reading the self-guided tour plaques with their usual curiosity. There's no sign of them, so we carry the canoe to the designated drop-off location before finding our way to the Eldorado Hotel. My dad had left a message on the satellite phone saying that he'd reserved a room large enough for all of us to stay.

We check into the hotel, take showers, and have just walked outside again when we see my parents pull up, their little yellow dog grinning out the back window. Always up for an adventure, my parents are almost caricatures of good sports. They're game to hike up a mountain or tour on a crowded city bus. They'll sleep on the floor or on a plane without complaint. Retirement for them is less about slowing down than speeding up; driving a day or two down the bumpy Dempster Highway to meet us in Dawson seemed like a perfect summertime outing. And in the back of all of our minds is the knowledge that such travel might not always be possible. Though my parents have both been fit and healthy most of their lives, we've learned how quickly this can change.

With a big squeeze for each of us, my mom tells us we look great, ignoring the fact that we have changed back into the dirty clothes we've been wearing for the past week. My dad waits his turn for a hug and says that we must have made good time on the river to already be here and showered. He's tanned and smiling, and I'm relieved to see that he seems to have changed little since the spring. Still, I can't help but notice that his shoulders look thin and hunched. His tremor is apparent. I swallow a wave of sadness and try to push my worries aside.

For months after his diagnosis, we danced around the fact of my dad's Parkinson's. He alternated between a graciousness that humbled me and a stubbornness that made me want to strangle him. He demanded the same of himself as he always had—up early, shoveling snow, washing my car unannounced, biking to work no matter the conditions outside. When his body refused to cooperate or it took a few minutes longer to accomplish the tasks that he had taken for granted, I'd hear him cursing under his breath. But he rarely unleashed his frustration on anyone but himself.

One afternoon, we sat down for coffee at a neighborhood café and I broached the subject of his illness for the first time. I only managed a few sentences before my reddening eyes gave me away. "I haven't asked you what is most scary about having Parkinson's," I said. "I know I don't always express myself, but if there's anything I can do to help, even just talking about it . . ." I trailed off.

He surprised me with the honesty of his response. "I've been a little disappointed that things seem to have gotten worse faster than I'd hoped," he said. "But the disease is different for everyone."

"It must be tough," I offered lamely. "You're so active, and still a hundred times more than the average person, and I'm sure it's not easy to change your expectations."

"My biggest worry is that I'll become a burden," he said. By this point I was crying openly. "Oh, honey, I don't want you to worry about me. I'll be fine. Plus, I really don't have anything to complain about. I have a wonderful wife and three wonderful children. There are a lot of worse things that could happen to a person."

Yes. And no, I thought then. For a man who has climbed the highest mountain in North America and biked across Europe, who routinely signed up for a running race in every town he visited, who squeezes more activity into long summer evenings than anyone else I know, a disease that threatens to rob him of his mobility could be devastating. At times, more than he lets on to anyone else, I'm sure it already is. Since his diagnosis three years ago, he's had to stop running and skiing. He needs additional time to get ready for the day. Simple tasks are made complicated by a tremor and worsening fine-motor skills. But he's also refused to let it claim his life. And in witnessing his private struggles, I have learned just how tough he really is. A journey like ours takes persistence. Dealing gracefully with a chronic illness takes an ocean of strength. I hope that somewhere inside of me, I have a bit of that same strength.

We help my parents carry their luggage inside—one tiny backpack for each of them and a large cooler for Pat and me. Knowing exactly what we would be craving most, they brought homemade cookies, cheese, fruit, vegetables from their garden, and beer. Almost everything else in the car is for us, too: boxes of food, hiking gear, mail, and a stack of books in case we need extra reading material. That evening, my parents treat us to a salmon barbecue and ice cream. It's a perfect meal for the hot summer day, the real food a welcome change after a week of gorging on rolls and snacks. Back at the hotel, we begin to sort our gear for the next leg.

A rare moment of calm during our journey up the Inside Passage.

Blue mussel bed revealed at low tide. Life abounds on the coast.

Battling stormy spring weather. Rain, hail, and near-freezing temperatures were common conditions during March and April.

A quiet cove in Princess Royal Channel.

Rowing past a sea lion haul-out.

The remote log cabin we built near Haines, Alaska. We had a brief respite here before continuing our trip in the Coast Mountains.

View from our cabin. The prominent notch is the mountain pass we must cross to reach the Yukon.

Skiing over the pass that is visible from our cabin.

Searching for a route back onto the ice. A glacier's terminus is often broken and steep.

Life on ice. The sun sets on our camp near the Alaska-Canada border.

Traversing around a glacial lake in the Coast Mountains.

Ascending a steep ridge in the Tombstone Mountains.

Watching swans during spring migration in Tagish Lake.

Enjoying the sun after a bath in the Peel River.

Wearing a head net to defend against the mosquitoes, Pat retreats from the quicksand-like mud of the Mackenzie Delta.

Moose wading in the Arctic Ocean are silhouetted by the midnight sun.

Crossing an ice bridge on the Arctic Coast.

Hiking along the shore of the Beaufort Sea, Ivvavik National Park.

Dragging our packrafts to deeper water. Near major river outlets, the Beaufort Sea can be shallow for more than a mile offshore.

Rough-legged hawk chick after its nest slid down a collapsed mud bluff on Herschel Island.

The nest of a common eider on an Arctic barrier island.

Cooking over an open fire. For most of our trip, we left our camp stove behind to save weight.

Backbone of a beluga whale.

Paddling through ice floes in the Arctic Ocean.

Hiking into the vastness of the Brooks Range. These remote mountains lie entirely above the Arctic Circle.

Musk ox skull on the Arctic Coastal Plain.

We followed caribou trails for hundreds of miles across mountains and tundra. Headwaters of the Hulahula River, Arctic National Wildlife Refuge.

Creek crossings were cold and frequent. Near the Continental Divide in the eastern Brooks Range.

Navigating snow-covered boulders on our final descent into the Noatak Valley.

Turned back by a late August snowstorm, we search for a route over the mountains.

On day 4 without food, we wait in the tent for our resupply to arrive. Upper Noatak River.

After following caribou for many miles, we are rewarded by the sight of the western Arctic caribou herd crossing the Noatak River on their fall migration.

Our first backcountry adventure as a family of three. Hiking with ten-week-old Huxley on the Lost Coast near Yakutat, Alaska.

As we pack, my parents click through photos on our digital camera. My mom, in her usual contagious enthusiasm, oohs and ahhs as my dad follows along on the map. When he sees a photo of an eagle, he eagerly reports that he spotted a king-fisher at the lagoon near their Anchorage house. "And the loon babies are getting big," he tells me. Over the last several weeks, I've gotten updates on the status of the loon chicks that hatched on a lake nearby their house. Though neither of my parents is an avid bird-watcher, they have done their best to tackle bird ID since I began working as an ornithologist. Their reports aren't always accurate, but they're delivered with such pride that I rarely correct them. I know the point isn't to check species off of a list but to honor their daughter's passion. Despite my somewhat unconventional career choice, my parents have never suggested I should study anything other than birds. They've never complained about the fact that I regularly return from the field smelling like rotten eggs, or that my academic pursuits have given way to tromping around in the woods. They've even managed to keep any doubts about our commitment to walk, ski, and paddle to the Arctic to themselves.

The next morning, my parents hike with us for the first two miles out of Dawson City. We follow a four-wheel drive track that takes us to a fire lookout station, where we'll leave the road and head cross-country into the Tombstone Mountains. It's a warm morning, and our packs feel relatively light. As we hug goodbye, I tell them that we intend to make it to the park headquarters in five days, possibly sooner. I run through the numbers in my head—fifteen miles a day, five days, no problem. My parents plan to spend a few nights at the park after dropping off our supplies, and we should see them again at the campground if we don't run into any major delays.

The moment we step off the dirt road and onto the soggy tundra, our feet begin to launch a protest. For months, they have been confined to rubber boots or hard-shelled ski boots, not the lightweight hiking shoes that now squish and slide on the wet ground. Between rowing, skiing, and canoeing, we've hardly walked at all. As we pound through the first half a dozen miles, it's obvious that the transition will be a painful one. It takes only a few hours of hiking for the entire sole of each of my feet to turn into a blister. There's no need to decide whether to drain the fluid—the blisters ooze and weep as I walk, and soon the open sores are packed with dirt.

From a distance, the Tombstone Mountains looked like excellent hiking. But instead of the dry ridges and easy trail miles we imagined, we've found a latticework of alders, wild roses that claw at our clothes and skin, and muddy, rutted tracks that lead to river crossings impassable at spring melt. It's the middle of June, but summer has not yet arrived. Each time we climb high to avoid the brush, we meet thigh-deep drifts of last winter's snowfall. The granulated snow scrapes our calves and shins and tears at our feet. When I lose my shoe several feet beneath the surface, I dig with bare hands as tiny balls of ice fill my sleeves. On the descent, we punch through to the unstable scree beneath and use our trekking poles as poor substitutes for ice axes when the surface becomes firm and slippery. In the valleys, creeks are flowing fast and high. We're making miserable time.

As we struggle through the tough terrain, my emotions begin to match my body's dwindling reserves. The worry and grief about my dad's illness that I tried so hard to push aside worm their way back in. I can't stop thinking about what this disease might take from him, and from us all. On the third day I hike only three miles before telling Pat that I need to sit down and rest. I'm light-headed and have been fighting a growing exhaustion that I can't blame on the tough travel alone. Following behind Pat, struggling to keep up even through the least heinous of the bushwhacking, I feel a lump building in my throat. This is the first real hiking we've done so far and I'm a mess. What is *wrong* with me? What will I do if I can't continue?

Later, Pat will tell me all he could think of was the exchange we'd had at the visitor center shortly before leaving Dawson City. The woman staffing the front desk was dressed in "uniform"—a Gold Rush–era dress and flowery hat covering long, curly auburn hair swept back into a silver barrette. She hardly looked the part of an experienced backcountry traveler but surprised us by offering to pull out maps of the area after we gave a brief synopsis of our planned route. As we pored over the contours of the Tombstone Mountains and the drainages that would lead us to the Wind River, she pointed out dozens of places she'd been to, including one valley where she and her husband spent an entire winter in a wall tent. As we finally turned to leave, she said, "Be careful. I'd watch yourself in there." Expecting the usual warnings about bears and hypothermia, I nodded and promised to be cautious. She continued, "No, I'm serious. Me and three of my friends got pregnant in the Hart River drainage. There's something about that place. My husband and I planned to do a lot more exploring but we had our daughter instead."

When my period comes a day later, I'm washed with relief. But once I know I'm not pregnant, there is no good explanation for my fatigue. Is the tough travel alone enough to make me dizzy and nauseous? Or is this all in my head?

I distract myself by thinking about my parents waiting at the campground, only a day or two away. Even now, at thirty-three, there are times when I could really use some childhood pampering. This is one of them. I know he's out of range, but

I decide to call my dad's cell phone anyway. Before we left, I had suggested that he check his messages from the park headquarters. On his voicemail, I try hard to sound cheerful rather than pleading. I explain that we're running late but will be there in a day or two and would really love to see them again.

The next evening, we're only two miles from the Dempster Highway when I vote for setting up camp. I'm no longer suffering from the mysterious bug that plagued me for several days, but my feet feel like they've been run through a meat grinder and I'm ready for a rest. "What difference does it make if we get there in the evening or the morning?" I whine to Pat. His feet are sore and blistered, too, and it takes little convincing before he agrees to stop for the night.

The next morning we wake early and make our way to the road, paralleling it for several miles to the Tombstone Park headquarters and campground. The hard ground pounds our raw soles even rawer. When I see the park sign, I look around hopefully for my parents. We don't see them or their car, so we hobble past the parking lot, leave our packs on the porch, and walk into the visitor center. When I ask the woman at the front desk about the resupply boxes, she says, "Oh yes, they're downstairs. And your parents are such lovely people. We really enjoyed them. They just left this morning."

She sees my face and continues, "Oh, I'm sorry, were you expecting them to be here still? We offered to let them use the phone, but they said they didn't want to bother us." I mumble thanks and walk outside, embarrassed to show my childish tears to the ranger or to Pat. My parents went home, we have our resupply boxes, and everything is fine. But I'm crushed. Just a couple of hours ago, while we hiked the last two miles to the road, my parents drove past. I won't see them again for months.

"Here, they left this for you, too." The ranger walks outside with a note. As I crunch on one of the two apples they included with our boxes, I read in my dad's handwriting, *Caroline and Pat, We're sorry we missed you. This is a beautiful place, it's easy to see why you love being out here so much. You two are amazing. Love, Mom and Dad.* Each of them had signed their name—my dad's tidy engineer's script next to my mom's sprawling cursive. My mom will tell me later that each day at the park they hiked up a trail going in the direction they expected us to come from. Before they turned around, my dad would call out my name in a half yodel, half shout. "Caaroooliiine!"

Suddenly, the prospect of being pregnant no longer seems like the worst thing in the world. Missing my mom and dad does. If parenthood inspires the sort of bond I feel with them right now, even from a distance, maybe my sister *is* right. Maybe having a child matters more than battling brush and postholing through last season's snow. Maybe family trumps wilderness. Or perhaps these pieces—made of illness and love and birth and death—are inextricably linked, tangled and messy like the green stalks of alder that grow on every hillside.

WIND RIVER

For the first time in a thousand miles, I've returned to a place I know. It's summer solstice, almost ten years to the day since we first laid eyes on the Wind River. Ask me to draw a picture of this riverbank and I wouldn't know where to begin, but plop me down here from anywhere in the world and I can tell you with certainty that *yes, I have been here before.* We are camped at the swirling confluence of two rivers—the Wind and its junior tributary, the Little Wind. Twenty minutes ago, paddling toward this confluence, I might have been twenty-three again, each detail newly etched in my mind. The steep canyon walls, the murky river, the peculiar tan-colored shoreline with sand like finely ground pepper. And Pat's eager, childlike grin as he bounced down the wave train where the two rivers join.

We're less than a hundred miles from the Arctic Circle, and the sun sits stubbornly above the adjacent red and brown cliffs, following a shallow arc across the sky. We won't see darkness for another two months. Canada's northern Yukon is a land of extremes, bruisingly cold one moment and unbearably hot the next. For most of the afternoon we paddled our rafts down the Little Wind River, gazing at a kaleidoscope of orange and slate-colored pebbles as they swirled beneath us. Just before the confluence of the two rivers, we drifted past a seven-foot-tall shelf of ice, a carryover from the winter's deep freeze, and I shivered as we passed through the refrigerated air. When we reached the larger river, we were flushed abruptly into its muddied flow, cold clear water twisting into whirlpools and disappearing beneath a silt-laden skin. We arrived at our island campsite to find hordes of mosquitoes gathered onshore, where the air hangs still and heavy. We sweat in the evening heat and I imagine pressing my palms against the river ice.

As we set up camp, I think about our earlier canoe-building trip. Back then, we were two kids on this river shaping a boat from spruce bark. Living in a wilderness to ourselves. For a time, this was all that mattered.

Now, a decade later, we have found our way back. We are no longer kids, no longer tentative about a future together, but our need for adventure is every bit as intense. I wake up each morning with Pat by my side, wondering less about where our relationship might take us than about what we will find around the next bend. On the first trip, when everything between us was new and fragile, the river seemed more a backdrop for our relationship than a necessity. Now I can see that it's so much more. Wilderness has become the silent third partner in our marriage, and here is where it all began.

After a meal of bland pasta doused in oil and salt, we lie naked in the tent, too hot for sleep, and help each other remember. Pat falls back on his sleeping pad laughing as I prod him about the horse. The fucking horse. The horse that almost made me quit. Several days after leaving the town of Mayo, we came across a dead pack horse from the previous fall lying in a dry aspen forest. The large, dark body stood out grotesquely against the mossy ground, its distended belly and empty eye sockets covered with flies. We paused to stare at the horse and Pat asked with the inquisitiveness that I alternately love *and* dread, "Why not a skin boat?" We hadn't yet found any birch trees large enough for building a canoe and were beginning to wonder if our plan might be doomed. In the carcass, which screamed to me of bad omens and bears ready to pounce, Pat saw only opportunity.

Thus began the biggest fight of our courtship. I told Pat that he could continue without me if he wanted to build a boat out of a dead, rotting horse in bear country. What kind of lunatic had I committed to spending a summer in the woods with, anyway? He listened quietly to my tirade, then, undaunted, pointed out the benefits of the horse skin. It was here, in front of us, and we might not find any birch trees. The book we carried with us described construction of skin boats—how different would it be to build a canoe out of horse hide rather than moose hide? He looked at me and said, with a completely serious face, that he saw no reason why we couldn't make it work. Only after I repeated my promise to leave, this time with hot, angry tears, did Pat begin to relent. In recalling this story, there's a part of me that wishes I were the one who could see opportunity in the most unlikely places. Pat has always allowed a vision to carry him where it will, even to the inside of a bloated horse.

The next morning, we pack up in a swarm of mosquitoes so suffocating that we forgo our usual breakfast routine and grab a granola bar on the river instead. We navigate the rapids we first paddled in our spruce canoe, our rubber rafts bouncing easily off of gravel bars as we scan for familiar landmarks.

"There's the cave where we waited out that rainstorm," Pat says.

"And I think that might be the pool where we caught so many grayling," I continue.

As we zoom downstream to meet the Peel River, a thunderstorm rolls through, pelting us with hail and sending light-ning flashing in electric streaks across the dark sky. By late afternoon, all traces of the storm have vanished and the sun warms our bodies. We strip down to T-shirts in the sudden heat. Just ahead are the Peel Canyon rapids, with their stand-ing wave train. Like an amusement park ride, five-foot-high waves stack one after the other in quick succession.

On the earlier trip, we had paddled through these rapids desperately, with little room for error in our makeshift canoe. A pair of red-tailed hawks circled high above us, their shrieks capturing the emotion of the moment in a single, searing sound bite. One of the birds followed us through the rapids until we reached calm water, its shadow dancing on the steep rock walls. There is a photo of me from the last trip, tri-umphant after the waves. In it, I'm backlit by the orange glow of the canyon, my hand-carved paddle raised above my head and the muscles of my back flexing beneath my black sports bra. In another, Pat's face is upturned to the sky as he bal-ances his paddle on his chin in playful celebration.

Today, in our rafts, we bob through without worry. When I hear a hawk cry overhead, I look up to see it launch from the burnished canyon wall, just as the other birds had a decade earlier. Red-tailed hawks can live two dozen years or more, often returning to the same nest site each summer. But life is dangerous, even for a large raptor, and it's likely another pair has come to take the place of the birds we saw from our ca-noe. Still, I let myself believe that today's hawk has been here all these years, standing watch, waiting to usher us through the canyon safely once again. After exiting the rapids, we re-lax and let the current carry us downstream. The river is wide and fast, and there's little navigation required. I dip my hand into the water. It's surprisingly warm, and I have a sudden in-spiration. "Pat, will you raft up next to me so I can swim?" I ask him. By now we're both sweating in the heat, itchy from the bug bites of the recent days, and long overdue for a bath.

He raises his eyebrows and shrugs. "I guess we'd be fine if we took turns."

As I strip off my clothes and slip into the water, I feel like I've been baptized. The water courses over my sticky body, washing away more than dirt and sweat. When I climb out and let the sun dry my skin, I can hardly keep my eyes open with the immediate, intense pleasure. Seeing my blissful response, Pat decides to follow suit. After our baths, we dunk our clothes into the river, using the current to wash away the superficial layers of filth.

The sun caresses my breasts and warms my belly and thighs, and I suddenly remember the delicious feeling of bare flesh. This newfound clean has awakened an intensity of desire that has been masked alternately by bugs and mud and cold and fatigue. I glance over at Pat's nakedness and briefly contemplate crawling into his boat. But the river is still moving quickly here and neither of us can recall exactly what lies ahead. Instead, I bide my time, searching for a good campsite, whiling away the hours until we stop for the evening. When we finally choose an island to land, we climb onto the bank and sink into knee-deep mud. Onshore, the mosquitoes are horrific and we scramble to pull on our clothes and head nets. I cook the fastest meal we have—instant ramen—and we crawl inside the tent to eat.

We finish only half of our dinner before we're naked again. We don't need words to tell each other that we're thinking the same thing; our baths have left us hungry for much more than food. Here, on this first, formative river, our desire is larger than life. We reach for each other greedily, oblivious to the mud and bugs that now coat our skin. For this moment, the world exists only inside our tent, our bodies moving in a rhythm that is both familiar and entirely new.

PART FOUR
Arctic Coast

MACKENZIE DELTA

The world is gray. Flat. The odor of rot hangs thickly in the air. Water swirls and rushes and goes nowhere all at once. Wind blows us backward but does nothing to deter the mosquitoes. At each meandering bend of the river, mud crumbles into the silty water. Like the riverbank, I, too, am crumbling.

I squint through a cloud of mosquitoes as my raft bumps against the standing reeds that line the channel. *Quit. Quitter. Quitting.* Trying these words out in my mouth, I feel each one as it passes. I scoop water heavy with sediment into my bottle and try to swallow without tasting, ignoring the silt as it grinds against my teeth, the sour flavor of peat on my tongue. Pat lifts his head net to take a bite of granola bar and glances over at me. He's consumed by the task of transferring food to his mouth without exposing his face and neck to the giant swarm of biting insects that circle around his raft.

When we first reached the southern edge of the Mackenzie Delta, eight days after leaving the confluence of the Wind and Peel Rivers, the bugs had seemed more of an annoyance than a hazard. Now, these tiny creatures define our days. My hands are scabbed with bites scratched bloody and raw, my left eye is swollen partially shut, my ears are ringing. It doesn't seem possible that, more than weeping blisters or avalanches or twenty-foot seas, mosquitoes could destroy us. But they have discovered all of our weaknesses, finding their way down our shirts and up our pants, between the mesh of our shoes, through fabric where it presses closely against our skin. As they bite and buzz, we take turns losing our composure. I cry, then scream, then give up and do the only thing I can: paddle like hell. Pat's lowest moments appear in the form of long, angry silences or questions about whether we have made a terrible mistake by choosing a route through the delta. At these times, I play the optimist, however forced it might be. I say out loud that we *will* make it to the coast, but some days this feels like an outright lie. This is how we've kept ourselves moving forward, leapfrogging in shifts past each other's misery. But today neither of us can muster up even an ounce of positive thought.

On paper, we have a simple mission—to get ourselves from the Wind River in the northern Yukon to the shore of the Arctic Ocean. Conveniently, the Mackenzie River flows north, in exactly the direction we want to go. The catch is that this river ends in one of North America's largest deltas. Here, at its mouth, the Mackenzie Delta stretches more than fifty miles across. Easily visible from space, its enormous web drains one-fifth of Canada's land mass. On the ground, this translates into mosquitoes, mud, and stagnant water. It's a labyrinth of confused channels and willow-clad shores.

As we began planning our route more than a year ago, we knew enough to dread the Mackenzie Delta. We had heard that the river was sluggish, the bugs bad. But even after all of our combined years in Alaska's backcountry, we had failed to appreciate what bad really looks like. When we passed through the village of Aklavik several days ago and mentioned our plan to paddle to the coast, people stared blankly at us. They were stunned by our stupidity. No local would be caught dead paddling at two miles per hour through this part of the delta during summer. Anyone in the area who can afford to takes a *motorboat* to the coast, where the bugs are driven away by the wind.

Even the caribou are smarter than us. They, too, escape to the coast or to the tops of windy ridges during the peak of the mosquito season. If they don't, they're harassed to the point of madness, or worse. Caribou biologists have estimated that mosquitoes can drain up to ten ounces, equivalent to an average cup of coffee, from a single animal in a twenty-four-hour period. This translates into a daily barrage of sixty thousand mosquito bites. At such intensity, anecdotal reports of calves dying from blood loss by mosquitoes hardly seem exaggerated. In fact, for a brief annual period in the Arctic, the biomass of mosquitoes outweighs that of caribou. Given a pound-per-pound conversion of 20 million mosquitoes per caribou and herds that number in the tens or hundreds of thousands, the calculation is staggering. But my pockmarked body doesn't need math to recognize the consequences. The only caribou on the lower Mackenzie Delta right now are those with a death wish. And the only humans here are us.

For all of the horrors of this place, it has one redeeming feature: the ducks. The Mackenzie Delta is one of the most productive breeding areas for waterfowl in the world. At least that's what I've been told. I wouldn't know it from being here, where I can see little besides the inside of my head net and the steep-sided, brushy banks of the endless sloughs we must navigate. I try to remind myself that, for some creatures, this isn't hell on earth.

After Pat shoves the last of his granola bar beneath his

head net, we pause to check our location. He points at a squiggle on the map, and I count sixteen mosquitoes on his hand. When we look up again a minute later, we have floated a hundred yards upriver. The combination of wind off of the Arctic Ocean and high water levels has made the current flow in reverse, pushing us south instead of north. We're on a giant liquid treadmill with no apparent end.

I hate this place. I hate myself for being broken by bugs and mud and wind. "I don't want to waste another second of my life with this bullshit," I say to Pat. "I want to go home."

Pat doesn't reply, but I can see his face, pinched with concern. The instant I open my mouth I am speaking taboo. Of all the topics we've covered in our thousands of hours together on this trip, neither of us has given voice to quitting. It's too tempting an offer to resist—a plate of brownies placed in front of a dieter; a bottle of Ronrico offered to a recovering alcoholic; a shower and four walls provided to me right now. The line between words and actions quickly becomes blurred, so we don't usually speak of our deepest doubts, or acknowledge that we could cause our own undoing. Today I don't care. I can think of no reason to continue, no reason to be here. All of the images that for months have made me smile—photos of my chubby little nephew, flight paths of migrating birds, the memory of a breaching whale—only remind me of all the other places I'd rather be.

But no matter how much I want to quit, there's no easy way out of this situation. The nearest town is a hundred miles upriver. Our next destination is Herschel Island, a tiny territorial park that happens to be the northernmost point of land on this stretch of coastline and also one of the most remote. It's impossible to leave the delta without soliciting a rescue. Even pleading the threat of blood loss or insanity or both, I don't think mosquito harassment qualifies as an emergency.

So I'm free to entertain the possibility of giving up without accepting its consequences. Pat is kind enough not to point out the obvious when I tell him that I'm ready to end our trip. Sometimes deciding on the impossible is cathartic, and today I'll take whatever relief I can get.

We paddle until a growing headwind slows our already meager pace in the packrafts to almost nothing. I check the GPS and see that we are clocking in at 0.1 miles per hour. At this rate, it would take us another four hundred hours to reach the coast, so we decide to head to shore. I pull up next to Pat on the bank, step out with one foot, and sink to my thigh in muck that smells like a mixture of feces and wet grass. We crawl to a thick stand of willows, where we might be able to escape the worst of the mud. But the tangle of woody stalks is hardly an ideal campsite; as soon as we have one tent stake secured, the other side of the tent floor pops up again. Eventually we trample the bushes into a semblance of flat ground and begin the hurried shuffle to get our gear and ourselves inside with as few mosquitoes as possible.

The smell in the tent is overwhelming. Despite being covered with dried silt, sweat, and squished bugs, we haven't made any attempt to clean ourselves in more than a week. My toenails are stained brown from the tannins of the soil, the adjacent skin pimpled with bites and angry pink scars. When I zip the tent door closed, we begin our hunt. Slap, slap, slap. Mosquitoes seek shelter in low places; we shake our sleeping bags and sweep the corners of the tent to roust them. When we're finished, the relief is hypnotic. I'm on a reverse high—no stimulant, no comfort or discomfort, nothing except simply being. The delta has reduced me to a reptilian state.

It's 10:30 p.m., and besides the subtle change in the angle of the sun, there's little that distinguishes this time of day from any other. Normally, I love the Arctic's round-the-clock daylight. But here it's merely a reminder of how easily each hour blurs into the next. Meals, which typically define our days, have gone the way of every other luxury. Spending time among the mosquitoes to find dry wood—already scarce in this land of mud and bushes—and build a fire is unthinkable. Tonight, dinner consists of a bag of trail mix, cold instant black beans, and a chunk of cheddar cheese. Before adding the dehydrated beans to the plastic quart-sized yogurt container that doubles as my food dish and bail bucket, I use the corner of my T-shirt to remove remnants of grass and dirt, followed by a rinse of Mackenzie sludge water.

When I wake in the night to pee, I reach for the tent zipper until I remember where I am. The entire mesh door is covered with mosquitoes. The tent quivers with their buzzing. I simply can't do it. I won't bare my butt to the mosquitoes again. I have a terrible and liberating idea. *Urine is fairly sterile, right?* As I squat over the dish, then dump the pee out the tent door, allowing only two mosquitoes inside in the process, I can't believe it didn't occur to me sooner. I have no shame anymore.

In the morning, I rinse my bowl with river water before adding dried oatmeal and cold water. Granules of instant coffee stick in my teeth. Everything about this meal is disgusting, but I'm just glad not to be out with the bugs.

Two days later we're finally within a dozen river miles of the Arctic Ocean. Even as the bushes become smaller and sparser throughout the morning, I work to keep my expectations low. The ocean still feels far away. Several hours after leaving camp, the channel we're following spills into a series of lakes. We paddle across one of the lakes to its apparent outlet only to find that, once again, the water is flowing the wrong way. We're desperate to stretch our legs and anything seems better than the excruciatingly slow progress of our rafts. The bank looks fairly continuous, so I do a quick experiment. I hop out, shoulder my pack, and begin walking. My boat drags easily behind me and the ground feels surprisingly solid underfoot. I start to gain on Pat as he struggles against the current. Once I pass him, I celebrate my results. Walking *is* faster than paddling. Soon he joins me and we hopscotch along—walking when we can, then jumping into our rafts to cross the dozens of remaining ponds and sloughs. We practice this in-and-out routine for hours, passing a large flock of molting, flightless white-fronted geese whose movements resemble our own. Paddle, waddle, paddle, waddle, and repeat.

In the early evening, we finally reach the end of land, where a low sod bank folds into the silty water of the Arctic Ocean. The difference between here and the Mackenzie Delta from a day ago is staggering. The coastal breeze keeps the mosquitoes at bay. Without head nets blocking our vision, we can see again. Without bugs hovering around our noses and mouths, we can breathe. At our feet is a carpet of tiny, pink saxifrage flowers; above us, the sky opens into a million shades of blue. We notice a string of white dots on the horizon and, as we draw near, the specks become swans, floating serenely in the still water. I spin in place, scanning with my binoculars, and begin to count. Ten, twenty, sixty swans scattered across the flats. Phalaropes swim circles in the small ponds, while dowitchers probe, sewing-machine-like, along their margins. A Lapland longspur chortles sweet notes as a sandhill crane walks past with lurching, exaggerated steps, parading its prehistoric grace. This is the Arctic I had imagined.

We continue to hike across the tundra for several hours, heading west as we parallel the coast rather than the river.

From the air, this landscape resembles a strangely mowed golf course, pale green polygons as far as the eye can see. Permafrost creates this montage; ice wedges intrude into the soil, gradually increasing with each cycle of freeze and thaw. Miniature drainage canals are formed, and these eventually give way to ponds and sloughs. The wet ground is covered with tussocks, the massive, aboveground root systems of *Eriophorum* cotton grass, which look lovely covered with white puffy blooms. In practice, these plants are the bane of the Arctic hiker, creating wobbly, ankle-spraining mounds that are separated just far enough to make hopping from one to the other almost impossible. But compared to the trials of the Mackenzie Delta, even tussocks seem pleasant today.

We move slowly, stopping to watch a jaeger flash its long forked tail or a gyrfalcon float overhead. Each time, Pat's careful eye picks out a detail that I had missed. *Wow, look at those feet. The color of its eyes is amazing. They fly like they're swimming underwater.* Sharing observations is one of the gifts of traveling as a pair; together we see much more than I ever would alone.

I lift my binoculars to scan for more birds and notice a dark shape parked near the coast. In the low-angle sun, I'm not certain what I'm seeing. But then the hump moves. It takes only a second more for me to recognize that it's a large grizzly pawing at something on the ground, fur rippling like wind on water. I lower the binoculars just as the bear lifts its head, and I realize I've been fooled. I'm so habituated to using binoculars to magnify distant objects that when I gaze through glass my brain tells me that the bear must be far away.

Not so this time. The bear is close. And big. And suddenly running toward us.

As the distance between us and the bear shrinks, we respond just as we've been trained—stand tall, wave our arms, and shout. The bear doesn't seem to notice. We're moments away from impact. When it stops a dozen yards from us, rising on its hind legs to sniff the air, I can make out the upturned edges of its nostrils. We stand frozen, arms raised like scarecrows. The bear faces us for several time-arrested seconds. The green grass glows. The breeze is cool against my sun-warmed cheek. The raspy calls of cranes echo from somewhere in the distance. I can feel, more than see, Pat standing tensely next to me. And then, just as suddenly as the bear charged toward us, it turns, glancing back only once before launching into a full-fledged run across the tundra and out of sight.

Pat and I stare at each other, attempting to digest the fact that we narrowly avoided being plowed down by a grizzly. Before the panic has had time to settle, an explosion of jet-black clouds billows toward us. Lightning flashes across the sky, and we jump when a resounding crack of thunder follows. The first drops of cold water pepper our faces as whitecaps froth on the ocean surface. We race to set up our tent next to the shelter of an enormous driftwood log, close to where we had first seen the bear. For two hours we hunker down, listen to the rain pounding against the tent, and press our palms against the nylon walls to buffer the wind.

Then, almost as quickly as it had appeared, the storm breaks. What remains is a glorious fresh world. The grasses shimmer with beads of water, the ocean has stilled completely, and the last streaks of lightning share the sky with the sun and the moon. The scare of the bear has passed. We climb out of the tent, ready for a midnight meal. Pat scouts for a place to make a fire as I dig a bag of pasta from my pack. I hear the surprise in Pat's voice before he explains what he's found: a partial carcass of a bearded seal, matted down in the tundra. There are fresh diggings all around it; the grass has been turned over and bits of discolored flesh lie scattered nearby. The bear that had nearly flattened us was not feeding on grass, as we had thought, but on this rotting seal. And here we are, camped right next to an ursine dinner buffet.

Before leaving on our trip, we had been warned about the bears along the Arctic Coast—these barren-ground grizzlies are reputed to be hungrier and more daring than their larger, salmon-fed counterparts to the south. Like so much of life in the Arctic, potential grizzly bear prey, such as caribou calves, emerge only in brief pulses. An opportunity for seal meat would not be easily overlooked, and it's obvious that we can't stay here.

We pack up the tent, inflate our rafts, and slide down a grassy slope into the water. The sea is calm and I fall easily into the rhythmic motion of paddling. After the birds, the bear, the storm, and now the golden reflection of the sun on the sea, my senses are overloaded. It's magical and disorienting at once. I feel like I'm watching the night unfold through a television screen. Even the seemingly impossible feat of reaching the coast seems abstract and distant, as though it has happened to someone else. I've left any thoughts of quitting behind with the mud and the bugs.

Soon, we round a point and a pair of big brown objects come into view. They're far offshore and seem to be floating on top of the water. As I squint through my binoculars, I can begin to make out the details. Broad chest, long nose, humped back and... antlers. I am staring at two moose wading in the ocean! Moose are common in ponds, sloughs, lakes, and rivers farther south. But moose in salt water? And in the Arctic Ocean no less?

Here, at the northern periphery of the continent, the

ocean has a daily tidal range of barely a foot, and large underwater deltas extend from river outlets that deliver silt and debris. The effect of this is a coastline that's oddly shallow, sometimes measuring only three feet deep even a mile from shore. I can see now that the moose are standing upright, with all but their lower legs exposed. We attempt to work our way around the wading moose, our paddles bumping the bottom with every other stroke. First we try to pass on the inside, near shore, but reconsider when they take several curious steps in our direction. These long-legged animals can move much faster in the shallows than we can paddle, so we offer them a wide berth, taking a detour by heading farther offshore. But then the moose change direction, too. At first their advances are almost imperceptible, just a couple of steps every few seconds. On land I wouldn't think much of it; moose generally want little to do with humans. Pat and I joke about how outrageous a tale of a moose attack in the Arctic Ocean would be. But when the moose start to gain on us our laughter dissipates.

"What the hell?" I say to Pat as I hit the mud with my paddle yet again. Even when we reach deeper water, the moose seem undeterred and begin to swim behind our rafts. Soon, all we can see are their antlers and backs, and we put our heads down and paddle hard. They follow us for nearly a mile, their noses stirring the surface of the water as they stare blankly ahead. Only after Pat and I are sweating and scared do the moose begin to slow their pace. When they finally give us a break from our all-out sprint, we're left guessing about their motives. Moose are relative newcomers here, where warmer temperatures have promoted growth of willows and other shrubs that they like to eat. Maybe these moose were simply curious, our presence undoubtedly a much rarer occurrence than theirs. Or maybe, as implausible as it had first seemed, they had in fact been aggressive. At the end of a very strange day, anything seems possible.

As we paddle west into the sun's glare, adrenaline is quickly replaced by fatigue. I look down at my watch. It's 2 a.m. We choose a place to land and stumble through the mud to shore. We drop our packs, hurriedly set up camp, and coerce a few muddy sticks into flame. Even though it's long past midnight, this will be our first proper dinner in a week. After we finish our chores, we collapse onto the moist ground. Silently, we devour a pot of pasta, flavored with smoke from the cooking fire, and squint into the sun.

After a few minutes, Pat points out a glaucous gull perched on a driftwood log a few hundred feet away. He tells me with a straight face that the bird is a giant satellite dish from one of the Cold War–era Distant Early Warning (DEW) Line sites. These conspicuous white geodesic domes are visible from dozens of miles away and have been characteristic features of the Arctic landscape since their construction in the 1950s. Most are now abandoned, and provide homes for nesting falcons and families of foxes rather than military defense. But the object Pat's looking at is most definitely not a satellite dish. I glance at his expression to see if he's joking. He isn't. Just about anywhere else in the world, I'd be concerned by his response. Instead, I point out the now-airborne gull and laugh. In this landscape of flat and flatter, it's shockingly easy to mistake a two-and-a-half-pound bird for a giant, golf-ball-shaped station that measures sixty feet in diameter. With few recognizable landmarks in the foreground, our eyes struggle to make sense of size or distance. This is the visual trick of the Arctic. Nothing is ever as it appears. As we lean back against a driftwood log in a landscape devoid of trees and gaze out at a sky curved under its own enormity, it seems entirely possible that we've reached the end of the earth.

CAPSIZED

D oes the Arctic Ocean count as a real ocean?" I ask Pat. I don't mean technically. I won't argue with a named body of water. But functionally it's unlike any other ocean I can imagine. Ice-choked for most of the year. Shallow for miles offshore. Barely discernible tides. Not much salt. A flat gray horizon. Perhaps, among lakes, the Salton Sea could claim an equal measure of weirdness. But here there is also endless daylight in summer, endless darkness in winter. There are a hundred kinds of ice but no trees. It's a landscape that feels scant pressure to conform.

Before we left, we weren't sure whether we'd be hiking or boating this coastline. We could find little information to guide us. We could find little information at all. But packrafts allow us the luxury of not knowing. Each day we can choose whether to walk or paddle. Since we reached the coast, we've done equal parts of both. This morning, there's only a light easterly breeze, and we decide to inflate our boats to take advantage of the tailwind. In the sunshine, the small waves seem benign, and we watch for belugas as we paddle toward deeper water. Flocks of male eiders and long-tailed ducks stream by, their contrasting black and white plumage flashing against the water. The shore grows increasingly distant, but I think I can make out several caribou grazing on the tundra. I'm happy to be here today, on this strange ocean, in a world that will always be foreign to me.

It takes only half an hour for us to realize we have made a mistake. Even before we've noticed a shift in the wind, the waves begin to stack higher and tighter. The water changes suddenly from gray to green as we drop off an underwater shelf into the main outlet of a nearby river. Our many weeks spent rowing are not much use to us here—the rules of the two oceans share little in common. The wind blows chop toward land as the river pumps fresh water out to sea. Our rafts bend and bob in the confusion.

I tighten the straps holding my pack onto the bow of my raft and secure the few loose items in my cockpit. Cold seeps through the raft's thin rubber hull as seawater presses against my legs. I try to move from my core to conserve energy, but soon I'm paddling any way I can to stay afloat. We've worked ourselves more than a mile offshore and we must somehow make it back. We angle our boats in the direction of land, waiting for breaks in the larger sets. One stretch of shoreline shimmers white; chunks of sea ice have stacked up along a narrow strip of beach. I try to ignore the nagging fact that the mud bluffs ahead look to plummet directly into the surf. Will we be able to get out even if we make it to land? Or will we be ejected by steep cliffs and jumbled ice? *Shh,* I say to myself out loud, *Just shut up and paddle.* I focus on the task at hand, pausing only long enough to relieve some of the tension building in my muscles and joints. I flex and unflex my hands, shrug my shoulders, and will my cramped hips to relax. The wind is blasting now, sucking the heat from my torso and blowing strands of hair into my eyes.

Between waves, I sneak glances at Pat, paddling twenty

feet to my right. Each time he looks more confident than I feel, riding high over the crests of the waves, shooting me an occasional smile of encouragement. We travel this way, side by side, for what feels like hours. When an especially large set rolls through, I panic as my paddle twists underwater and threatens to pry me out of my raft. Just as I manage to find my balance again, I look to Pat for reassurance. Instead I catch his eyes as they widen in surprise, then shock.

Parallel to the surf, he's caught in the trough of a five-foot wave. He braces to the left, leveraging his paddle against the water to stay upright. A few seconds later, another wave passes under him and he braces to the right. This time, he can't get out of it. In slow motion, he lists over until water begins to pour into his boat. Then he topples into the icy sea.

As he goes under, I feel the air leave my lungs. All I can see now is the bottom of his raft.

I struggle to turn into the wind and toward Pat. I manage to spin around just in time to see him pop up, paddle in hand. Water dribbles from the brim of his hat, surprisingly still attached to his head, and sends tiny rivulets down his cheeks. He grabs the corner of his boat and his blue eyes shine brightly against the dark seawater. He looks very much alive. And focused. "I'm OK," he tells me. "Dammit, I can't believe I flipped."

But the ice-studded water is barely forty degrees and he needs to get back into his boat quickly or he'll become hypothermic. That's not easy to do with the raft upside down, bobbing wildly in the waves. I alternately paddle and brace, working my way toward Pat. Each time I gain ground, the wind pushes me back. Finally, between gusts, I paddle frantically and cover most of the short distance between us. As I come near, Pat manages to right the raft on his own, heaving the weight of his pack up and over its buoyant tube in a motion similar to flipping a tortoise onto its back. I help to steady his boat as he climbs in, belly first. He spins around to face the waves. His teeth are chattering slightly, but he's burning with the adrenaline of being dumped. His cheeks are flushed. He's still alert and strong.

With a quick glance back to make sure I'm close behind, Pat angles toward shore, his stroke fast and aggressive. The surf breaks white and foamy below the bluffs that are now only several hundred yards away. The beach is narrow, but there isn't any ice here. Pat shouts that since he can't get much wetter he may as well go first. I hold back and watch. He surfs in, staying upright, then half carries, half drags his boat out of the water, sprints back down the beach, and motions for me to follow. Waiting for a large set to pass, I see a gap and make my move. As I land, a wave crashes into my cockpit, soaking me from the waist down. Pat grabs me by the wrist and pulls me onto shore.

We stand stunned for a minute before reverting to the obvious need—get warm. The cold is beginning to set in for real. My fingers refuse to cooperate when I try to unroll my dry bag full of clothes. Pat struggles with wet pants that cling to his skin. Somehow, despite my hands, I manage to snap a few photos. In them, Pat is shirtless and wearing an expression so intense that anywhere else it might have seemed staged. Eventually, we wrestle our soggy clothes off of our bodies, wring them out, and change into dry layers. As we roll up our rafts and repack our bags, we exchange more raised eyebrows than words. What is there to say after moments like these? Most of what I'm thinking demands too much of us to speak aloud. *I'm glad you didn't die. I'm glad I didn't die. Thank God we're on shore. What if? What if...*

* * *

Before we left on this trip, Pat's mom, whose pragmatism often overwhelms her sentimentality, told me bluntly that she was most worried about disaster befalling only one of us. At first I missed Joanne's point and assumed she was voicing typical motherly concerns, exercising the instinct of keeping her children safe and healthy. But then I realized that she wasn't talking about her own potential for loss, but ours. She couldn't imagine how either one of us could live with the trauma of losing the other, especially by a plan of our own devising. Though I had a ready answer for most of the criticisms or questions about our trip, this one was harder to refute.

Ever since the early days of our relationship, I've struggled with the idea of being left behind. When Pat and I first lived together in Bellingham, after our summer on the Wind River, this fact nearly swallowed us. As a new graduate student I was often too busy to join him on local adventures, and he would head out without me to climb a frozen waterfall or try a new mountaineering objective. He was bold and fearless in the mountains and saw little need to justify his actions. I tried to embrace the same qualities in him that I first fell in love with, but his solo adventures left me confused and scared. Previously, we'd bonded over making decisions as a team. But when we were apart, the hazards of the mountains loomed large and terrifying. Suddenly I was the one at home waiting for news, and I hated it.

I quickly learned that it was easier to be mad than scared, so I made myself sick with worry, then cursed at the emptiness. As I cooked dinner on our tiny sailboat and counted the hours until Pat would return, I pushed away thoughts of avalanches and bloody falls. When he finally arrived, I picked fights over grocery bills and whose turn it was to clean the mildew from our sailboat's hatches. After I had worked through my list of petty complaints, the subject inevitably turned to what levels of risk were acceptable to each of us. He didn't think he was doing anything particularly risky, he told me. I reminded him of all of the accidents that had happened to people doing the exact same thing. Even the strongest climbers have been injured or killed, often leaving a partner, or an entire family, behind. It wasn't a matter of trusting his competence; I didn't want to sign up for the certainty of loss.

For many months we boomeranged between planning a future together and tipping toward a breaking point: On our worst days, Pat felt stifled; I felt abandoned. He tried to ease my worries by calling more frequently to tell me he was OK. He left me with detailed route information and itineraries. I distracted myself with long trail runs and meals with friends. But only as the spring semester approached and we began to prepare for a summer of sailing and climbing did our relationship move beyond a state of limbo. Our solution for managing risk was to take it on together.

But today I can see clearly how Joanne was right. Being together is no guarantee of safety. And the bolder we get, the more we have at stake.

For the rest of the afternoon, we stay on land and follow the strip of beach between the ocean and the mud bluffs that rise a hundred feet to the flat coastal plain above. Along one especially narrow section of beach, I spot a family of five brant geese—two adults and three fuzzy young—waddling in front of us.

As we walk, the birds run ahead in an attempt to escape our unintended advances. The parents dart back and forth, frantically herding the youngsters with each nod of their charcoal heads. The goslings look like oversize lint balls with legs as they totter down the sand, riding the violent edge of the surf zone. It's tough terrain for baby geese only days out of the nest. They struggle over logs and wade awkwardly through soft sand, their feet clumsy beneath them. A peregrine falcon swoops down from the bluff and the parents scramble to shield their young, wings outstretched, hissing an empty warning. The family angles toward the water, but retreats at the sight of crashing waves.

Pat and I take a break on a log to eat a granola bar and discuss our options to get around the geese. It's obvious that the goslings are tiring quickly. If we keep pushing them, the adults may leave the youngsters behind, forcing them to decide which hazard is worse: the waves or a falcon's talons. But we can't climb the bluffs; the crumbling mud banks are too steep here. At their current pace, waiting out the geese could take days.

We decide to attempt a pass. If we can get ahead of the birds, they'll presumably stop their frantic dash. We hoist our packs onto our shoulders and begin to jog down the beach. At a high mud impasse, we find two of the goslings already swishing helplessly in the surf. When a large wave pushes them above the waterline they stumble onto the sand. The adults and the largest gosling have continued running and are almost out of sight. A glaucous gull and a rough-legged hawk have joined the peregrine overhead, all jockeying for an easy meal.

Now we're left with two choices: return the abandoned goslings to their parents or watch them become dinner. Pat defers everything bird-related to my court and looks to me for an answer. I shout that we should try to catch them as I begin to chase the smaller one and quickly corner it in a pile of driftwood. Pat grabs the other one as it attempts to scramble up the steep bluff and instead tumbles backward into his hands. I instruct him to tuck the bird's head under a sleeve to help keep it calm as we chase after the parents at as close to a sprint as we can manage with our packs and fragile cargo.

We catch up to the rest of the family just in time to watch them plunge into the surf. The lone gosling dives in behind its parents, somersaulting several times before making its way past the break zone. Now we have two baby geese in our custody and a wall of waves between us and their parents. "Oh shit, I think I screwed this one up," I say. Pat shrugs and suggests that we try to get them into the water somehow. We wade in as far as we dare and on the count of three toss the two goslings toward the others. The adults fly off when we approach, honking as they rise into the gray sky and disappear. We duck behind a log to watch, and I curse myself for thinking I was such a clever bird expert. We have pushed the parents away *and* sent the babies into the same rough seas that had ejected Pat just hours earlier. I can't imagine how I would explain this to a fellow biologist. I don't know if I could tell my sister that I drowned two baby geese. I feel ridiculous and devastated at once.

But then we hear honking overhead. I look up to see the adults circling, arching their necks in search of their goslings. When they spot them in the water, their calls increase in volume before they land with a splash next to the young birds. My heart lifts. It's not a perfect solution, but with their parents nearby at least they stand a chance. Now that the goslings are away from shore, the raptors and gull have given up their hungry watch. Past the surf zone, they float easily in the swell. They need to last in the water only long enough for the wind to ease and the waves to lessen. Already, it seems much calmer than this morning.

We watch for several minutes more before continuing down the stormy coast. For the rest of the afternoon, I stare out at the sea and hope that the goslings will end their water voyage like we did—shaken but safe.

A LAND OF CHANGE

Eight days after reaching the Arctic Ocean, we arrive at Herschel Island, a small piece of real estate with a big history. Twenty-one degrees of latitude and four months now separate us from Bellingham. It feels like a lifetime has passed. When we'd arranged for a food resupply to be flown in by a park ranger on a shift change, Herschel Island had been just a teardrop on the satellite map, a tiny intrusion of land that seemed an improbable destination. In our frantic preparations to leave, I had focused only on checking the tasks off my list. *Pack food that will keep for many months without refrigeration. Contact park rangers. Fill out the customs forms.* I hadn't imagined what it would be like to actually stand here, a continent on one side, a sea on the other.

Flanked by two long spits that extend like gravel streamers toward the mainland, Herschel Island juts abruptly into the Beaufort Sea. For most of the year, this landscape is frozen solid. Just two weeks before our July 12 arrival, the last of the pack ice surrounding the island gave way to open water. Now only a smattering of ice floes remains. Pat and I rest against the lush tundra as we watch a pair of semipalmated plovers tend noisily to their new chicks. The fist-sized birds scurry back and forth across the gravel, shiny black necklaces flashing in the sun. In a cartoonish gait that shorebirds have perfected, their slender legs pedal rapidly as their bodies stay still and upright in an avian rendition of Wile E. Coyote's wheeling appendages.

The plovers' current mission is to keep us away from their chicks. One parent crouches and scolds me as the other feigns a broken wing, a distraction behavior meant to draw predators away from the vulnerable, flightless chicks. Plovers engage in what we call "biparental care," which is a scientific way of saying that mom and dad work together to raise the kids. A male plover swaps incubation duties with his partner regularly, allowing her plenty of time to feed throughout the day. He also sticks around to finish tending chicks after the female leaves on her southward migration. There are a number of reasons why such behaviors make sense—incubation is energetically costly, and thus it's more efficient to share the burden; females must replenish their reserves after egg production. But evolutionary theory and objectivity aside, these traits are also charming. Among birds, I'd much rather be married to a loyal, egalitarian plover than a solitary eider, who enters in a bustle of showy plumage and stays only long enough to copulate.

As we watch, the three chicks rush to hide beneath the feathered skirt of the parent with the dragging wing. They attempt to squeeze into a space far too small for them, stubby wings and tiny beaks poking out in every direction. It's ridiculous, and ridiculously adorable. Male and female plovers look identical so I can't tell them apart, but I make a guess anyway. "Good job, Papa."

Warm weather has seduced us into forgetting, at least briefly, the ticking clock of winter. As we relax in the sunshine wearing only T-shirts and shorts, I pointedly ignore the fact that Kotzebue is still twelve hundred miles away. The gravel beach holds a thousand little worry beads beneath my soles as I run my bare toes across the smooth pebbles. We're taking advantage of the warmth to air out our feet, which haven't been dry inside our shoes for a single full day since we left several months ago.

Once the birds scurry over a rise and out of sight, we lace up our shoes and walk toward a cluster of buildings to introduce ourselves to the local park rangers. Lee John, the Inuvialuit ranger on staff, welcomes us with a handshake and begins to tell us about the island. He was born half a century ago on Komakuk Beach, a remote stretch of coastline forty miles east of Herschel Island. Growing up, he tells us, his wardrobe contained a utilitarian combination of furs and commercially available fabrics. His mother sewed sealskin boots lined with felt and pieced synthetic quilting into a caribou parka. Each summer, his family came to the island where he now works.

Despite the fact that it's staggeringly remote and surrounded by ice most of the year, Herschel Island has been a centerpiece of change in this region. Ancestors of the Inuvialuit used the area for hunting and seasonal residence, and the oldest structures on the island date back more than a thousand years. But there is little visual evidence of these original sod dwellings. Instead, we see a tidy arrangement of clapboard-sided buildings painted white with red trim, starkly out of place among the island's weathered driftwood and muted tundra.

In the late nineteenth century, the quest for baleen and blubber arrived in the Arctic. Herschel Island offers a rare natural harbor, the only one along the entire northern coastline of Alaska, which allowed whaling ships to overwinter safely. Upon discovery of this protected anchorage, the island was transformed, almost overnight, to a bustling whaling hub complete with missionaries, the Royal Canadian Mounted Police, sports teams, and grand balls. The sudden flurry of activity also drew Inuvialuit, Inupiaq, and Gwich'in people from across the western Arctic; they sometimes came more than a thousand miles on foot or by dogsled to supply furs, meat, and labor to the whalers. The demand for natural resources quickly took its toll on the local environment and its people. Following a ninety-thousand-year residence, musk oxen were extirpated as a result of overhunting. They have returned only as a result of successful reintroduction efforts. Many indigenous residents died from venereal diseases, measles, and influenza. Their traditional ways were challenged by newly imported materials, including alcohol, and the pressures of a cash economy.

The whaling industry crashed just two decades later, a result of plummeting prices and declining whale populations, and outside interests soon shifted to trading furs and other commodities. The Hudson's Bay Company established a post here in 1915, selling supplies to local people and acquiring furs to be transported to southern markets. But even this period was short-lived. By the time Lee John was born, mercantile activities at Herschel Island had ended and the lone Royal Canadian Mounted Police outpost was closed. Since then, the island has returned to its formerly quiet state, now a far-flung territorial park that draws only a small number of intrepid foreign travelers and Inuvialuit hunters from surrounding villages.

We walk through the old buildings, listening to Lee John describe his mother's meetings with white traders when she was a child, and I try to imagine, in the not-so-distant-past, this island as a very different place. Fuzzy black-and-white photographs of Herschel Island show ice-fast ships, whaling crews bundled in furs, and hand-lashed dog sleds. The low shack that once held baleen later became housing for employees of the trading companies, then a summer camp for Lee John's family, and, eventually, a storage shed for the park. More than a century after commercial whaling ended, this remote island is destined once again to become part of the global economy. As sea ice melts, the Northwest Passage is being transformed from an impenetrable frozen maze to an open waterway. Projections of ice-free passage are only decades away, and countries are already vying for their share of this northern transport corridor. Given its unique geography and protected anchorage, Herschel Island will undoubtedly feature prominently in the "new" Arctic.

In the afternoon, we collect our resupply box, sort food and gear for tomorrow's departure, and head out for an easy hike along the narrow beach north of Pauline Cove. We pick our way along the bank as the mud bluffs grow taller and steeper, studying how the island's frozen soils are sliced clean by the sea. Underneath a thin layer of peat, giant ice lenses, several feet thick, protrude from the bank like the inner layers of an ice-cream cake. Soon the beach narrows and we're skirting frigid water, our toes slowly numbing in the cold. Before we begin to scramble to drier ground, we notice a pair of pale forms resting against the dark gray mud. Perched just a few feet above the waterline are two downy, baseball-sized chicks. I assume the chicks are dead and prod the larger one with my finger. It surprises me by returning a piercing yellow gaze. Its smaller sibling lies motionless, eyes closed, and I watch intently for its shallow breaths. We look up at the bluffs overhead—a scar of wet mud traces the path that these young birds followed as they slid downhill.

Over the previous several days, we've counted dozens of peregrine falcons, gyrfalcons, and rough-legged hawks, their nests perched precariously on mud bluffs that overhang the beach. Unlike the intricate weavings of songbirds' nests or the gaudy dwellings of ravens, the nests we've seen typically consist of little more than a scrape in the mud or a depression in a patch of tundra. Our ears have become attuned to the raptors' hoarse calls as they sound alarms of intruders in their neighborhoods. One morning, after a pair of gyrfalcons woke us at 3 a.m., we unzipped our tent to find fresh bear tracks in the sand.

Now, as we crouch over the young birds lying in the mud, the silence is audible. There are no shrieks or swoops or angry cries. I scan the skyline for their parents. Nothing. To an un-trained eye, many raptor chicks look alike, the unfeathered skin around their eyes distinctly reptilian, the beaks small and gray. I haven't encountered enough of these youngsters to know the difference, but a headless lemming lying nearby offers a clue—falcons hunt birds; rough-legged hawks feed themselves on small rodents. These must be hawk chicks. Even though a meal lies within reach, the young birds aren't likely to survive. The icy water threatens to swamp them in high winds. Foxes, martens, or bears might amble by. And they may have gotten injured in what must have been a wild ride down the eighty-foot drop.

Despite years of working as a scientist, well trained in the canon of objectivity, I want nothing more than to pick up the chicks and carry them back to the ranger's house, where they might be raised by hand. The larger bird watches me, big eyes peering from beneath a tufted, fuzzy head. It has wings that look far too large for its body, a yellow gape around its mouth that turns up into the semblance of a smile. The sum of these awkward parts is comical and endearing. But it's not their cuteness or the fact that they don't stand a chance in this world that is hardest to bear as I turn away from the chicks. It's the fact that they are helpless and I am not.

One of the first tenets of biology is to let nature take its course. It's also one of the hardest to obey. During my first field season studying kittiwakes in Prince William Sound, I blatantly violated this rule. As I visited the nests each week to track the growth of the chicks, I quickly realized that life at a seabird colony is a necessarily brutal affair. With the hatch of thousands of eggs came more death than I'd ever seen. I hadn't intended to be a kittiwake foster mother, but when a gray ball of fuzz waddled over and nestled itself between my boots one day, I saw no other choice. Its parents had disappeared and the chick pinballed between nests, pecked and harassed by every other bird it encountered. A returning field technician partnered with me that day mentioned casually that the crew had a history of raising one or two especially pathetic chicks each season. This was all the justification I needed. I wrapped the bird in a towel, brought it back to camp, and fed it by hand until it fledged from the back porch.

Now I appreciate the difference between wildlife rehabilitation and field biology. I understand that the needs of a species can't be addressed by tending to a single individual. For my graduate research, I captured chickadees and brought them into a lab, knowing full well that I would never be able to release them again. I did this because I wanted to help *all* chickadees, to figure out why so many birds suffered from a mysterious and debilitating disease. Still, with more than a decade of science behind me, I question each act that values knowledge over compassion. The nineteen-year-old who raised a kittiwake in the corner of a log cabin always hovers nearby. And each time she emerges, I'm secretly glad to see her.

Today, with the hawks, my feelings are more complicated. Behind simple pity or kindness is the knowledge that every one of us, including me, is partially to blame. We each hold a share of responsibility for the collapsing bluffs and warming permafrost, for the fact that this island is slipping, chunk by chunk, into the sea. But I also know there's little I can do for the birds. Even if I wanted to help, I have no cabin to raise them in, no lemmings to offer as food. I whisper a small good-bye before we walk away. I don't let myself look back.

As we return from our hike, we pass a stream of stocky black birds with red feet that glow in the sun. The birds pour in and out of the windows of an old church building that once hosted regular services for the island's residents and now houses this colony of black guillemots. They are not the only guillemots that have found a home in an artificial setting. I tell Pat about another colony on Cooper Island, across the border in Alaska, where the birds nest in hard-shell waterproof suitcases normally used to transport electronics. Concerned about the growing threats of polar bears—pushed to land as melting sea ice retreats farther offshore—an eccentric biologist named George Divoky created a village of bear-proof boxes.

Divoky first stumbled onto an unlikely seabird outpost on Cooper Island thirty years ago, when a handful of pioneering birds that normally nest in rock crevices much farther south determined that old navy debris on a small Arctic island would do just fine. Drawn to the easily accessible nests and novel location, Divoky studied guillemot breeding biology. A decade or two later, it became clear that these birds would teach him much more. Their small colony offered a dramatic look at a changing climate. Temperatures warmed, sea ice melted, storms intensified, Arctic cod became more difficult for the birds to find, and, eventually, polar bears began to visit. One season, bears destroyed all but a single nest, and Divoky hatched the idea of bear-proof boxes. At a bird biology conference in Anchorage several years ago, I watched video footage from a remote camera placed inside a nest box. A giant black nose snuffles at the plastic while a bird sits on her eggs, safe inside. As the video clip played, Divoky stood near the podium and grinned. Though at first glance the boxes seem outlandish and driven more by a man's attachment to "his" birds than a tenable conservation strategy, unconventional solutions may be just what we need in these unconventional times. Hard-shell suitcases won't help the fact that the entire Arctic marine ecosystem is in peril, but at least the guillemots of Cooper Island have one less thing to worry about.

When we get back to camp, we show photos of the abandoned chicks to Lee John. Shading the camera screen with his hand, he confirms that these are young rough-legged hawks. He tells us that Herschel Island boasts the highest breeding density of this species in the Arctic and that he's found several other nests that slid into the ocean. As warming temperatures melt the permafrost underlying the tundra, the coastline is subject to slumping at a massive scale. Some stretches of the island's shoreline are receding at nearly one hundred feet a year.

"There's not much we can do besides record the observations," he says. We're standing outside the ranger's cabin and his eyes scan the water while he talks to us. He points out a handful of gulls circling in the distance, a likely sign of belugas nearby. As we watch for whales, we learn about other recent wildlife casualties: caribou thin and ragged after winter, fewer able to return each year; a dozen musk oxen found dead along the coast, encased in ice. As Lee John speaks, I struggle to imagine weather so vicious that it could penetrate a musk ox's coat. Their long outer guard hairs reach nearly to the ground, swish around their bellies and legs, and create a microenvironment that can exceed ambient temperature by more than 150 degrees Fahrenheit. We saw a lone bull several days ago, standing motionless on a gravel spit, looking every bit as stoic as I had envisioned. It's not the biting cold but the unseasonable warmth that's deadly for these iconic animals. Recently, freeze-thaw cycles have become much more common during winter. Warm Chinook winds from the south bring melting and rain, followed by subzero temperatures that plaster the landscape in a sheet of ice. Animals are transformed overnight into living ice sculptures, their coats frozen solid against their bodies. These cycles are not unlike the ice storms of the northeastern United States, but here there's nowhere to hide.

Later that evening, we share a meal with several scientists who are studying changes in permafrost on Herschel Island. The impetus of the scientists' research, besides understanding more about climate change and frozen soil, is to identify a replacement site for the park headquarters and its historic buildings when the spit is flooded. Not if, but when. With current projections for warming and increased storm surge, the entire spit will be underwater in less than fifty years. For an island that sits on several feet of ice and is vulnerable to erosion on all sides, finding a suitable alternative is not a simple task. Here, historic preservation assumes an entirely different meaning as even the land is fated to disappear.

As a biologist working in Alaska, I'm no stranger to the effects of a warming climate—there are few topics that garner more attention. But my understanding is shifting. From the people we've met, the birds and bluffs and storms we've seen, I've learned a different set of facts. There's little local interest in discussing hypothetical consequences; the effects are evident here and now. The Arctic as we know it—a land of persistent ice and snow, a home to walruses and polar bears—is quickly becoming legend. The permanence borne of cold, the secrets locked in ten-thousand-year-old frozen soil, erode a little more each day. Despite the models and studies, none of us can predict exactly what is coming. It's an experiment with many variables but no standards or controls. There's only one thing we can be sure of: changes are under way, whether a person, or an entire government, chooses to believe in them or not.

Just days after we leave Herschel Island, heading toward the Alaskan border, we wake to a sea of ice. Overnight, a northern breeze has transported a city of ice floes from far beyond the horizon line to the coast. Suddenly, the idea of a warming climate seems laughable. It's freezing here! After the brief pulse of summer, winter is on its way again. Ice-cooled air gathers onshore, temperatures drop, and we progressively add more layers of clothing. Expansive views of the northern sky fade into raindrops and fog. We dig hats and gloves out of our packs and huddle around smoky fires before crawling into damp sleeping bags dreaming of sunshine and blue skies.

On the days we're not confined to land by ice too thick to navigate, we follow a series of slender spits and offshore islands that offer easy walking on firm sand and gravel. In aerial photographs, these barrier islands resemble long gravel snakes that parallel the mainland. Some are separated from one another by just a few hundred yards. Others are miles apart. Each time we reach the end of land, we inflate our rafts and paddle in the direction of what we think is the next island. We've learned that our map—last updated in 1956— no longer reflects the current realities of an eroding shoreline, so we launch into the gray abyss unsure of what we will find. The islands' ever-changing shapes are dictated by a constant game of tug-of-war. Massive rivers flow north and empty their contents into the sea. Meanwhile, storms and shifting ice pummel the islands' banks, which rise only inches above the waterline.

One misty morning, we follow Nunalik Spit to its end, listening to the eerie creaks and pops of melting ice. The moist air displaces sound, muffling our paddle strokes and magnifying noises from distant sources. Wind whistles through scoter wings as a flock passes high overhead. A pair of yellow-billed loons calls hoarsely back and forth. We eventually reach an adjacent island and hop out. As we roll up our boats, Pat notices a gentle ripple in the water fifty feet offshore. Then another ripple, and we catch the first glimpse of a bright white back. Belugas! We watch three shiny whales slice through gray water like inverse shadows. They circle twice before vanishing into the fog.

We begin to hike and soon stumble onto the skeletons of hundreds of other animals. Bleached white bones are everywhere underfoot, as if an entire museum collection had been emptied onto the beach. Like the driftwood that makes its way from tree-covered landscapes hundreds or thousands of miles away, these remains have been deposited by wind and currents. It's an Arctic natural history lesson, a study in oceanic transport, and a testament to mortality all at once.

We stop first at a jawbone from a bowhead whale. I turn the gently curved bone on its end, and its mouth is larger than I am tall. Pat bends over to pick up a vertebra and can barely lift it to his knees. Bowhead whales are not only enormous; they also hold the record for the longest-lived mammal on earth. The discovery of Victorian-era harpoon points in recently harvested animals attests to this fact, and scientists and Inupiaq elders estimate that some bowheads may live to be two hundred years old. Next to the heavy whale bones, the skeleton of a sandhill crane is airy and light. Its hollow wing bones, designed for flight, are ready to lift off in the wind.

Like the whales, the cranes offer a glimpse into time that far exceeds human notions of longevity. Cranes have resided on our planet for at least 2.5 million years, making them one of the oldest avian species. Despite our heavy-handed influence on the planet, we're comparatively short tenants here. Everywhere Pat and I turn we see more bones, and we try to guess at their origins. A caribou head balances upright on the sand, pointing its antlers toward the sky. There's a fingerlike structure that might be the flipper of a bearded seal. A short, tubular bone looks like the femur of a duck.

A few minutes later, a female common eider, sheltered behind a small gravel rise, explodes from our feet in a flurry of wings and feathers. These chunky sea ducks breed on barrier islands of the Arctic coast, their nests well hidden among piles of driftwood. We bend over to examine four olive green eggs nestled in a depression in the sand, and I finger the blanket of down that lines the nest, still warm from the heat of the bird's body.

The only other time I visited barrier islands on the Arctic coastline, I had come for the eiders. As part of a research team studying the breeding biology of this species, we spent a summer camped at a remote field site near the western border of the Arctic National Wildlife Refuge. During June and July, we traveled by boat each day to track the progress of the nesting eiders. But just when the chicks had started to hatch from their eggs, a storm ravaged the islands. Five days of driving rains and gale-force winds tossed sand along the shorelines and raised the normally indiscernible tide more than three feet.

When we were finally able to return to one of the islands, displaced eider hens wandered aimlessly, suddenly lacking the purpose that drove them to sit on eggs for twenty-three hours a day. Their nests had been washed away or buried under piles of sand, and the remains of newly hatched chicks—tiny beaks frozen in place, down shiny and wet—were strewn about like piles of driftwood. On all of the islands we encountered the same destruction. In just two days, nearly an entire breeding season was destroyed. This has always been a land of storms, but in recent years they have become much worse. New weather patterns create greater instabilities. More open water means bigger waves. Less sea ice means less protection from the surf. The barrier islands are the first to feel the impact.

The eider hen watches us from a distance, waddling slowly around a large driftwood log, biding her time until we leave. It's clear that there's only one thing she wants in the world right now, and that is to sit on her eggs. I can't do anything about the storms or the rapidly changing climate, but at least I can offer her this small courtesy. I cover the eggs with down to keep them warm and we hustle away from the nest, ducking out of sight to watch. A moment later, she's back, once again sitting still as stone. She'll wait here until her chicks pip from their eggs or a rogue wave washes over them. She'll wait here, trusting that a narrow island at the edge of the earth is exactly where she belongs. For all of its blowing sand and raging surf, this is home.

BARTER ISLAND

Ten days after we leave Herschel Island, we're nearing Kaktovik, a small Inupiaq village located on Barter Island where we'll pick up our next resupply. This is our last day on the coast before we plan to head south into the mountains. We have a tailwind and we're making good time, but the waves have grown large enough that we're surfing more than paddling, and my hands have started to ache from gripping my paddle. We pull ashore on a barrier island half a mile from Kaktovik to assess the final crossing. My eyes are scoured from the inescapable sun and wind of the Arctic summer. My nerves are raw from the difficult boating.

As I drag my raft along the sand, I notice two sets of bear tracks: one large, one small. We've seen enough bears by now that the tracks barely register as notable, even those of a sow and cub. Until it occurs to me a minute later that the only bears on a barrier island this far offshore are the big white ones that swim. When we look more closely we can see that the entire island is covered with paw prints. Huge ones. Fresh ones. Some look to be only a few hours old. The island is a polar bear highway. We had been warned about polar bears along this stretch of coastline, and we've kept our eyes and ears alert as we traveled west. On a foggy day, we would almost certainly bump noses before we'd be able to pick out a white bear in the mist. But this is the first time we've had real reason for concern.

I spin around several times, scanning the island. No bears in sight, but who knows when one might come slinking out of the ocean. Now we're faced with two rather unpleasant choices: confront the waves and hope we can make the crossing safely, or wait on the island for a bear to appear. Neither option is appealing but I cast my vote—*go now*. I'm not prepared for an entire night of waiting for an unwanted visitor. Pat pauses for only a moment before agreeing.

A yellow-billed loon is snorkeling near the island, apparently unbothered by the waves or the prospect of polar bears nearby. Unlike us, loons are built to swim. They have dense bones that can resist pressure underwater, fewer air sacs than other birds to reduce their buoyancy, and muscles designed for diving. But these features come at a cost. On the rare occasions when they need to move on land, they must propel themselves by repeatedly falling forward. Today, stuck on a barrier island with howling wind and sea spray, I'd gladly give up my bipedal speed for the bird's aquatic grace.

We launch into the rough water and paddle anxiously past the loon, willing every white wave to be only that. I would love to see a polar bear, but not from my raft. As we watch the shoreline of Barter Island grow closer, an odd structure comes into view. From a distance, it looks like the frame of an old Quonset hut, with nine large, curved beams arranged in a parallel fashion. Eventually, we can see that it's the giant rib cage of a whale. We will soon learn that this "bone pile"—a site where the remains of bowhead whales are deposited after subsistence hunts—is a permanent community fixture, a landmark referenced by locals when providing directions. It also serves as a polar bear snack bar. A unique truce has formed between the hungry bears and Kaktovik's human residents. If the bears are well-fed, they are generally no trouble. So the village offers up the remains of their hunt and in turn generates income from tourists who fly in from all over the world to see these enigmatic animals. Today, the bone pile is thankfully clear of bears; the bones were gnawed clean months ago and the supply won't be refreshed until the fall hunt.

We paddle to shore without incident—and without any sightings of polar bears. I'm relieved to be on land, near a community, even one that invites its local bears to dinner. We follow a gravel airstrip and then one of a handful of unpaved roads into town, walking past dilapidated buildings and modest plywood-sided homes. Several people nod to us, but unlike with other villages we've visited, no one smiles or says hello. The residents of Kaktovik are inured to tourists, like we are in Anchorage in the summer. To the locals, we're just two more tourists looking for polar bears.

After a full lap around town, which takes barely ten minutes, we come across a cluster of trailers patched haphazardly together with plywood, tin, and cardboard. A hand-painted sign adorned with a pair of wooden snowshoes reads "Waldo Arms." We walk in to find a combined restaurant, hotel, and convenience store that serves hot food from a kitchen window looking out onto a communal living room. The lodging is not only unconventional; it's also exorbitantly expensive. A room with a shared bathroom costs $225 per person. A shower is fifteen dollars. A load of laundry is twenty-five. In remote northern villages, costs of goods are proportional to the distance they must travel, often by a combination of jet, bush plane, boat, truck, and snowmobile. When we balk at the prices, the owners, Walt and Merilyn, offer us a reasonable rate in a back shed where we can spread out our gear and set up our tent on the floor, safely out of the reach of any wandering village bears. We follow a rickety boardwalk to the outbuilding, which leans steeply to one side. Inside is a jumble of broken engine parts and other discarded junk. In order to make space for our tent, we find a shovel and a broom and get to work. We collect rusted nails, move old mattresses, and stack rotted wooden beams. This shed would hardly be considered a livable accommodation by any other standards, but for tonight it will do. We finish sweeping the crooked floor, pitch our tent, and return to the main building to splurge on a hot meal.

Walt hollers out to us from the kitchen. "Whatcha want? Looks like you could use some chili and cheeseburgers. That's what I'm serving tonight, anyway." Pat and I debate for a moment, gazing at the prices—twelve dollars for a cheeseburger, seven for chili—before I tell Walt yes. He can't hear me and I have to yell his name before he looks up and nods. He plops chili from a can into two plastic bowls, passes them through the window, and points toward the microwave. Grunting, he walks to the freezer and retrieves a stack of frozen beef patties. He breaks off two burgers and tops each with a slice of pre-wrapped American cheese. As they sizzle on the grill he tells us how he came to Barter Island more than fifty years ago to fly supplies and crew back and forth from the remote DEW Line sites.

"Now I flip omelets and fry burgers. Not a bad retirement."

We sit down at the long family-style table and say hello to a local man sipping coffee from a Styrofoam cup.

"Where'd ya come from?" he asks us.

Since we arrived in the Arctic, we've heard this same question dozens of times, from children, storekeepers, village chiefs, and everyone in between. Only after we'd tried answering with "Anchorage," thinking that they were asking where we were from, and then "Washington," where we had started, did we get it. People didn't want to hear about invisible places a thousand miles away but where and how we had traveled on *their* map of the land. Of course. Herschel Island meant nothing to me until I had been there. The Little Wind River is a real place now. Kaktovik is a village, with a bone pile and a gruff man who serves chili from his living room. Once we learned how to answer this question in context—Aklavik, Blow River, the Little Moose Slough—we no longer got puzzled looks. Instead, people would offer up their own experiences in those places or stories passed along from a brother or a friend.

When he learns that our last stop was Herschel Island, the man's eyes light up. He tells us that his grandmother was born there and relatives he's never met still live on the Canadian side of the border. He continues, explaining that the Arctic was a different place when his great-grandparents trekked to Herschel Island by dogsled. I show him digital photos of Herschel Island and of the border marker we passed several days ago, a simple metal post pounded into an unremarkable patch of tundra. His tone sobers when he talks about the future. "I've never made it across the border, but I've heard everything is changing there, just like it is here. I'm hardly on the land anymore. The ice is bad. Gas is too expensive. I don't even know if my kids will be able to stay here, since there's nothing for them to do." When our dinners are served a few minutes later—chili and cheeseburgers for each of us—we share a meal that has landed here from another world.

Before leaving town the next morning, we visit the Kaktovik post office to pick up our resupply box. The post office sign sits at an awkward angle below the raised blue building, another casualty of the harsh climate and unstable, permafrost-laden soils. As we walk in, I notice dozens of blue and yellow sticky notes plastering the wall. It's a ready assortment of messages to slap onto the door when the postmaster steps away. Which apparently is often. "Went to clinic, be back later." "Frontier not returning today." "Counter open after mail is sorted." "A LOT of mail." "Maybe 2:30." "Went home. 640-1973. Only call if important." "Frontier CANCELED to-day." "Try again at 3:00." "Video training in progress. Please check back later." "Running errands." "Due in at 1:30." If you can imagine an excuse for why the post office might be closed, you will find it here. In remote northern villages, weather trumps all. The mail service is no exception, and everyone around here knows it. From postmaster to park ranger, subsistence hunter to grocery clerk, people in the Arc-tic live by the whims of the land.

I'm still grinning at the notes when the postmaster appears from behind the window—his long gray beard, ponytail, and tie-dyed shirt more reminiscent of Southern California in the 1970s than a tiny whaling village nestled at the edge of the Arctic Ocean. When we mention that we're expecting two boxes sent here by general delivery, he says, "Oh, I know just the ones. They've been cluttering my office for about three weeks." I start to pull out my driver's license to claim the package, and he laughs and says that if we take the boxes off his hands, he doesn't care who we are.

We begin to explain that the boxes sat here so long because we were traveling by foot and packraft before he interrupts. "You came along the coast? Oh, I bet you've seen some bar-rels." And he's right, we've seen dozens of fifty-five-gallon barrels, scattered on beaches and pushed up on the tundra. So rusted, they almost fade into the landscape, except for their height and sometimes uncanny resemblance to bears. The postmaster then launches into a long, animated expla-nation of how the barrels got here. In the early 1950s, the military flew in thousands of gallons of fuel to support the air force presence and subsequent construction of the DEW Line sites. Because they didn't have storage facilities for fuel, everything came in steel drums. Once the fuel was used, the empty barrels became trash. By the 1970s, Kaktovik had amassed tens of thousands of them that littered the island. The military decided to clean up the mess and hired local people to gather barrels and stack them on a beach near the runway where they would later be flown out. As he describes the process to us, the postmaster pulls out a notepad and begins to sketch the dimensions of the pile, amounting to an area larger than the size of the entire modern village, barrels two and three deep.

"And then," he says, eyes staring intensely at us from above his bushy, pointed beard, "a hundred-year storm hit, and—POOF!—they scattered like confetti. So now you know why there are barrels all over this coastline. Once something comes to the Arctic, it never leaves."

Sitting beneath the post office twenty minutes later, we open our resupply boxes to find the dried food we had packed months earlier, plus two bags of cookies from my mom and a note in her messy, scrawled handwriting. "Have fun and take good care of each other! Not too much longer before you'll be back. We can't wait to see you!" Pat and I lean against the crooked posts and dig into the first batch of cookies. The chocolate has turned white and chunks crumble drily onto my lap, but each bite is rich and delicious. The cookies and my mom's message remind me suddenly of home.

The thought of returning is both comforting and jarring. The end still feels so far away. But it's already late July, and summer is quickly fading. I'm excited to see friends and family; I'm also anxious about what life in Anchorage might bring. Before starting the trip, I had thought about what it would be like to be gone. I had worried about missing home or failing or being miserable. But I hadn't let my mind drift to the idea of coming back. It seemed too distant, too abstract.

Out here, our energy is consumed by the needs of the present—choosing the best route, counting down the minutes until snack time, glassing for bears and birds. This focus is a gift, a way to free my mind of cluttered thoughts and worries that have no easy solutions. But I can see that this clarity won't offer an obvious path through a world that dwells not in the sublime but in the mundane. It won't change the fact of endless tasks and schedules and never enough time in the day. It won't make my dad's illness go away or relieve my uncertainties about an office job in the city. So much will have to change if we decide to have a family. This is terrain I'm not quite ready to navigate. For now, I enjoy the simple pleasure of stale chocolate and peanut butter. I'll leave the rest for later.

After finishing most of the cookies, cramming ten days' worth of food into our packs, and leaving a message for my parents on their answering machine, we head out of town on a dirt road that leads us past the village dump. We follow a scattered trail of plastic and other trash that has been blown by the prevailing easterly winds or dragged away by scavenging bears or foxes. As the postmaster predicted, barrels dot the tundra in every direction. When we reach the south end of Barter Island, we will inflate our rafts and paddle across the narrow channel to the mainland. From here, we'll turn our backs on the Arctic Ocean and head south across the coastal plain and into the mountains of the Brooks Range.

PART FIVE

Brooks Range

ARCTIC PULSE

The past four months have been blessedly free of sensational headlines and late-breaking news. Broadcasts arrive by satellite phone, if at all, usually days late and distilled by family or friends into manageable sound bites. I feel no less a citizen of the world, but my world extends only as far as I can see. The information we need comes by way of shifting wind currents or looming rain clouds, while our fingers trace contour lines rather than political lines. As a result, it's easy to ignore the fact that we live in a divided world. Although borders exist everywhere around us—defining one country from another, federal jurisdiction from state ownership, national parkland from private property—there's little evidence of these boundaries on the ground. We've encountered no border patrol stations or survey markers, no "keep out" signs or trespassing warnings. In fact, the only official notice we've seen since we reached the Arctic was a week ago: a single metal post, crooked and riddled with bullet holes, with "CANADA" written on one side, "UNITED STATES OF AMERICA" on the other.

Before we left, I was surprised to get unusually human responses to my queries about how to legally transit international borders if arriving via glaciers or tundra. I hesitated to give my name when I called each of the customs offices, for fear that I would be told that remote crossings were prohibited. Instead, in each case, I spoke to real people with kind voices who told me essentially the same thing: just let us know when you get here.

After we stepped into Alaskan territory, signaled by the dilapidated post, I called the border contact on the satellite phone and reached an answering machine in Fairbanks. I left a message as I had been instructed. "Hello, my name is Caroline Van Hemert, and my husband, Patrick Farrell, and I have crossed the border into Alaska. Feel free to contact me with any questions, although we will have limited communication for the next several months. Thanks for your time." I never heard anything more. We might have been caribou or geese, passing unnoticed from one place to the next.

But today, it's impossible to ignore borders. We've just landed our boats on the shore of one of the most contested pieces of real estate in North America. Thousands of miles from our nation's capital, politics are writ large here. The Arctic National Wildlife Refuge is a place that almost everyone has heard about even if few, including us, have actually visited. For decades, it has been made famous by two things: caribou and oil. As I step out of my boat onto the muddy bank, I see neither of these.

I've wanted to come here for almost as long as I've known the refuge existed. The images I've seen are striking: brown bodies of caribou blanketing the tundra like the passenger pigeons that once filled the sky, tens of thousands of animals moving across the land in sinuous waves. Pregnant caribou cows dragging swollen bellies over mountain passes, across raging rivers, and through ice and snow to reach the calving grounds. Calves, lanky and wet, trotting across the tundra just hours after they're born. The Porcupine caribou herd has become a symbol of wildness; one of the last great migrations of large mammals left on earth. Offering abundant food, safety from predators, and a coastal breeze that keeps mosquitoes at bay, the calving grounds are both a caribou nursery and a place of refuge.

But for all the idyllic images, it's the divisiveness of this land that draws the most attention. Known politically as the 1002 area, its future is uncertain. Because oil sits beneath the ground, it's not just another stretch of tundra. It's one that's potentially worth a lot of money. As with most other remote places, its fate depends on decisions made far from here, in congressional halls and courtrooms, largely by people who have never set foot in the Arctic.

Ironically, many of the political debates in Washington, D.C., center not on what we stand to gain or what we might lose by allowing oil drilling in the refuge, but on the inherent qualities of the Arctic. Certain politicians would have us believe that the 1002 area is a barren, empty wasteland. Many others counter that it's one of the richest places on earth. There's some truth to be found in both versions. The Arctic is a land of contrasts. Light and dark. Abundance and scarcity. Lush green and frozen white. There are few places so defined by life. There are few places so desolate. Quiescence is followed by lavish excess. Anyone who lives here, or has ever traveled here, will tell you that this is true. There's little sense in arguing about it.

Closer to home, the disagreement is not just political, it's personal. Local villages are split over the best course of action. In Arctic Village, located on the south side of the Brooks Range, where the Porcupine caribou herd travels in the winter, most people oppose development. For communities that depend on subsistence, caribou are synonymous with life. The Gwich'in of Arctic Village refer to the calving grounds as *Iizhik Gwats'an Gwandaii Goodlit* (The Sacred Place Where Life Begins). In Kaktovik, the village nearest to the calving grounds, most people are in favor of drilling. With few economic opportunities available to them, jobs are precious. Many Inupiaq residents of Kaktovik, like the man we met at the Waldo Arms yesterday, see oil development as the only way forward. From either perspective, a culture is at stake.

There's no denying the fact that both caribou and oil have a long history here—ancestors of the Porcupine caribou herd witnessed the extinction of wooly mammoths and the rise of human settlements; the oil might have formed 100 million years ago. What's less clear is how, and under what terms, they will continue to coexist. As I roll up my raft and stare across the coastal plain, I see not a battleground but a quiet patch of tundra basking in the rare warmth of a summer afternoon.

We hoist our packs onto our backs and make it only a few hundred yards before we meet our first defender of the refuge. An adult parasitic jaeger dive-bombs me as its two gangly chicks dart across the tundra. The bird's long, forked tail streams behind it as it swoops from the sky with a warning call that sounds like a cross between an angry cat and an out-of-tune clarinet. A moment later, it comes again, shrieking as it grazes my pack. The third time, the jaeger means business. It hits the top of my hat with its outstretched feet. *Damn you*, it seems to say in its rage. *I won't let a whole season's worth of energy be derailed by clumsy human footsteps.* I hold my trekking pole to the sky, hoping the bird will attack the aluminum rod instead of me.

The jaeger has every right to be defensive. Raising young

in the Arctic is not an easy job. Breeding birds have an impossibly long to-do list: find a mate, select a nest site, lay eggs, keep the eggs warm for several weeks, deliver food to the chicks for several more. This must all fit into an impossibly short window of time in an extreme and unforgiving climate. When birds show up on the coastal plain, ice still covers the ponds. Before they leave, the first snow will likely fall. In late July, we've arrived on the cusp of change. Most birds are finishing their duties for the season, and the drive to build nests and raise young will soon be replaced by an equally desperate yearning to leave. Summer is being edged out by fall. There's an urgency to this place that's palpable. This jaeger, like the millions of other birds that come here to breed, is going about its usual frenetic affairs.

It doesn't take a degree in ornithology to appreciate the fact that this is a productive place for birds, even after the peak of the summer has passed. Jaegers are just one of several dozen species that regularly breed here. Many more use the area as a stopover, often arriving at this migratory crossroads from distant locales. When I scan the tundra with my binoculars, I see birds everywhere. Geese, temporarily flightless, wander in molting flocks; loons paddle the margins of lakes; gulls and jaegers hover above us. Unlike many of the birds, the caribou have already left. Only rutted tracks and frequent piles of scat give any indication that, just weeks before, fifty thousand animals covered this landscape. The caribou aren't headed to a land of warmth, but to one that offers at least a few trees for cover. After calving, the Porcupine herd travels back across the mountains, young animals in tow, to the forested southern flanks of the Brooks Range, where they can more easily find food to eat and shelter from the fierce winter blizzards.

Standing here now, it's difficult to imagine the refuge's coastal plain marred by oil platforms and the persistent scars of ice roads. The land feels as ancient as the caribou that use it. Lichens, which can persist for millennia, cover many tundra surfaces. Permafrost holds frozen secrets under the peat. Even the cotton grass plants are old, surviving a century and a half or more, their thin roots holding tight to the shallow soil. I try to picture a revised version of Arctic wilderness—caribou calves milling around shiny smokestacks and long-armed drills. Geese nesting next to a helicopter pad. A revolving community of shift workers traveling from as far away as the birds, here not to raise families, but to make a living to support them. No matter how small an environmental footprint oil companies promise with new technologies and more stringent guidelines, the impact of such activities is undeniable.

The jaeger interrupts my thoughts with a cry and another screaming descent toward my head. Visions of petroleum stocks and angry, red-faced politicians are dwarfed by this simplest of acts: a parent defending its child against the big bad wolves of the world.

Leaving the jaeger to care for its chicks, we parallel the Hulahula River, christened more than a century ago by Hawaiian whalers homesick for warmer locales. As we hike upriver, steep peaks rise before us with pale green slopes and anemic glaciers that seem misplaced in this land of persistent ice. Like everything else here, the mountains are unfamiliar, offering a stark contrast to the wet, lush slopes and broad ice fields of the Pacific Coast Ranges we know well. Despite the interesting views, we find ourselves staring constantly at the ground. After several months in the Arctic, our eyes have learned to read clues, however faint, that indicate the presence of an animal trail: grasses leaning sideways, purple stains of crushed crowberries bleeding into ivory-colored lichen, small branches that are bent and broken. Our feet can sense the slight increase in firmness when we're on route and perceive the springiness of the tundra when we're not.

Without trails, we would flounder and flail. This landscape is vast and unwieldy. Its riverbanks are too steep, its tussocks too large, its rock faces too crumbly to allow passage without some guidance. Here, where caribou reign, it's their tracks we follow. But unlike a human-made trail, which employs a single route by design, caribou have no such notions of exclusivity. Their trails wind and crisscross, stop and start. They make as many paths as the landscape allows. On flat, easy terrain, the animals fan out so widely that trails all but disappear. At natural constrictions, over mountain passes or along narrow river banks, one trail might funnel ten thousand animals. Here, crossing the tussocks of the coastal plain, we have a dozen options to choose from.

Coming over a small rise, we spot a lone caribou calf twenty yards away. At first glance, I see only its impossibly long legs and wide-eyed gaze, and I pull out the camera to take pictures. Then I notice the details: sharp protruding ribs and open sores where warble flies have burrowed under its skin. The animal nibbles on grass, but at this age it's clear it won't survive without its mother's rich milk. There are no other caribou in sight. As we stand and watch, the calf approaches us hopefully, tilting its head as it sniffs the air. Its message to us is obvious. "Are *you* my mother?" Somehow, in the flurry of motion that accompanied the post-breeding migration, this youngster has missed the action. Perhaps it became separated from the herd. Perhaps it was a runt, too small and weak to keep up with the masses. Perhaps it simply failed to pay attention. Whatever the reason, the mistake was fatal. Months from now, when we look at the photos, the calf will have starved or, more likely, been torn apart by a predator.

I can only hope the end will come fast. With a whisper of apology, we say goodbye and follow the rutted caribou tracks toward the mountains.

An hour later, we catch a glimpse of the ailing calf's future. On the tundra lies the carcass of another young caribou that has already become someone's dinner. The kill looks fresh, perhaps only hours old, and we search for signs of bears or wolves. But all I see when I scan the tundra is a rusty barrel in a small ravine. I point it out to Pat, who squints and says, "I don't know, I think that barrel has ears." Still incredulous, I pass my binoculars to him for a closer look. "Oh, that is definitely *not* a barrel." I realize we have been looking at two different objects—one a barrel and one a bear. As the bear rises and turns toward us, two small cubs following tightly behind, I feel my stomach tighten at the sight of the dead calf at our feet. The bears watch us until we reach the next bend in the river.

When we begin to climb into the foothills, the caribou trails we've been following taper into five tracks, then two; when the terrain gets steep and narrow, just a single path leads through the scree. It's well worn into the rubble, and the footing feels solid. We follow easily, thinking little about it until we begin to drop toward the river, first gradually, then sharply, until we have to slide on our butts to continue. The trail ends abruptly at the river's edge. The water isn't deep here, but it looks turbulent and cold.

Pat and I put on our warm jackets, grab a snack, and sit down to discuss our route.

"I can't imagine why caribou would go this way," I tell Pat. "Do you think it was a mistake? There must be an easier crossing."

"They usually seem to know where they're going," he replies. "But this might be the exception."

We decide to climb back up the slope and continue without crossing the river. At first the walking is easy, and we're smug in our decision. Thirty minutes later, our mistake becomes obvious. We reach a constriction where our only options are to climb up and over a mountain that towers several thousand feet above us, or drop into a steep-walled canyon with the river churning far below us. There's no way forward. We were deluded, however briefly, into thinking that human logic could trump ten thousand years of caribou intuition. We were wrong.

The caribou clearly know things we never will.

BROOKS RANGE
ANNIVERSARY

It's two days before our fourth wedding anniversary and I know I should stop with my complaints. I've made my point. I've been stewing all morning, ever since I returned from filling our water bottles at a nearby creek and caught Pat stuffing one of my food bags into his pack. "What do you think you're doing?" I demanded.

"You have too much weight," he lied, meaning instead, "I think we could move faster if I carried more of the load."

"This isn't just your trip, you know." Pat looks at me as if I have accused him of stealing an ice-cream cone from a five-year-old.

After a long pause he says in a measured tone, "Yeah, I realize that." And then he turns and continues along the rocky ridge we have been following for two hours, feet skimming over wispy puffs of mountain avens and the first reddening leaves of bearberry plants.

But I'm not ready to let it drop. I try again. "I want to do this myself. I'm not interested in being the 'wife' part of 'the adventurer and his wife.'"

Pat slows and I see a shift in his posture. Normally his long arms swing loosely at his sides, broad shoulders curving forward as though being pulled downhill, leaning slightly toward the next ridge or valley. Now he stands awkwardly erect and is gnawing on a narrow backpack strap that dangles by his right shoulder, a habit that usually emerges only when he is so physically exhausted that any distraction is welcome.

"Caroline," he says, addressing me by my full name, a salutation that feels oddly formal, one usually reserved for the public sphere. It's not his tone that I notice as much as the way the syllables of my name come out forced and cold. "Car-o-line." His nicknames for me had started as a joke—when Pat learned I had little fondness for rabbits, he started calling me Bunny. When my sometimes less-than-sweet tongue lashed out with sharp words, he'd switch to "Sugar." Others followed—"Peanut," "Bud," "Sweet Pea"— and what was once playful sarcasm has become second nature, so much so that its absence feels like a slap. He continues, "You're blowing this out of proportion. I was just trying to be helpful."

On many other backcountry adventures, I've been the only woman among a bunch of guys. Usually, I don't have any trouble keeping up. But add a heavy pack and the differences start to emerge, especially between Pat and me. Pat is built like a rugby player—all upper body. He develops muscles just by looking at a bench press. I, on the other hand, could have a lucrative second career as a burglar, with shoulders narrow enough to slip through even the smallest of window cracks. My sister, by no means a hulk herself, has taken to calling me "Small Back," a description that infuriatingly fits.

As minor an infraction as this morning's food theft may have been, it insulted my already fragile confidence. I'd like to let go of any pretense of proving myself, but humility is a hard lesson to embrace. It's not only on personal trips that this is true. Field biology often demands the same brute strength as carrying a week's worth of food on a backcountry traverse. Even though manual labor is only a part of my job, I'm acutely aware of the fact that I'll never sling a shotgun like it's a natural extension of my body or haul equipment with the ease of someone twice my size. Still, I want to shoulder my share of the load. I want to know that I've accomplished this trip myself.

But after reclaiming my food bags, I'm horrified to realize that I can't match Pat's pace. Each time I feel my pulse pounding in my head as I chase up a steep rise behind him, another surge of anger bubbles in my throat and finds its way to my lips in a tirade of self-pity. "You make me feel like you'd be better off with a different partner." Pat ignores my whining and offers his back in response. He picks up his pace, almost imperceptibly, and I fume. "Dammit, Pat, don't you even care enough to say anything?!"

We rarely argue, but today I'm prepped for battle. The recent stress, and subsequent relief, of swimming across the Chandalar River has faded into the monotony of placing one blistered foot in front of the other. Even without the weight of our packrafts, which we mailed from Arctic Village to Anaktuvuk Pass several days ago, travel has been slow and painful. We've stumbled over tussocks in the lowlands. We've scrambled along endless scree slopes. The persistent ache in my shoulders and the rawness of my hips where my pack has chafed away the skin have left me irritable and weepy.

We hike for an hour in silence interrupted only by cursory words related to route-finding, which we exchange like business associates rather than husband and wife. Despite my mood, it's impossible to ignore our surroundings. The sun is shining, the water of the adjacent creek runs perfectly clear, and a light breeze whisks away the heat. Willow leaves have started to darken and curl at their edges, the evening frosts signaling a change of seasons. Heather on the adjacent slopes blushes toward fall. My thoughts turn over memories of recent days—the Chandalar River, the gray-headed chickadees, the wolverine we spotted loping across a valley—and gratitude slowly edges out my anger.

After another thirty minutes of ignoring each other, I'm ready to call a truce. I've found my pace again; the cramps in my shoulders have loosened and I'm no longer struggling to keep up. Besides, the day is simply too beautiful to stay mad. I attempt to brush off the morning's argument, but Pat wields his silence like a sword.

"Oh, wow, it's a goshawk." Pat glances up to see where I'm pointing but says nothing.

"What else did your mom say when you talked to her in Arctic Village?" I ask, trying to squeeze a few words out of the still air.

"Everyone seemed fine," he replies. They had chatted for half an hour, so I know there must be more news, but this is all I get for now.

"Do you remember the name of the next pass?"

"No." As we walk, I trail a dozen feet behind Pat, offering him solitude as a form of apology. But when he stops to pull a stick from his shoe, I'm gazing up at the hillside and accidentally bump my knee into his hunched back.

"Sorry!" I exclaim. Suddenly the concept of personal space in a landscape so large strikes me as ridiculous and I choke out a guffaw. Pat turns to see what I am laughing about. I shrug, then lift my arms to the sky, beginning to giggle. His eyes soften as he suppresses a smile.

In the early years of our relationship, I would have held fast to my stubbornness, insisting that I was right to start an argument over what could be considered kindness. I would have demanded an apology before admitting that our partnership couldn't be measured in percentages or pounds. During our first summer together on the Wind River, I came up with the idea that, proportional to our weights, I should carry 40 percent of the shared load and Pat should carry 60 percent. Our food rations were divided exactly the same way. Sixty-forty. An easy calculation. Fairness. On this trip, our loads have been similarly balanced. But Pat has helped me realize that the point isn't equity. It's the act of helping each other. Despite my insecurities, I know the success of our journey doesn't hinge on strength alone. There's so much more. Route planning. Calorie calculations. Customs forms. Food preparation. The behind-the-scenes details that could freeze us in our tracks if done poorly, details that have mostly been my responsibility. We're in this together.

I'm suddenly ashamed that, just two days after the scare of nearly losing Pat to the Chandalar River, I've let my ego take over. The lessons that have come from sharing so many miles of wilderness matter much more than whether or not I can carry a few extra pounds and still walk as quickly as my husband.

At the top of the next ridge, we gaze down into a broad valley. I spot a pale shadow as it darts across the tundra at the bottom of the slope. "Look, Pat." I point, but he is already staring in the same direction.

"There's a second," he replies a moment later. Two wolves are working their way uphill, and I train my binoculars on them for a better look as we duck behind a boulder to watch. The animals zigzag back and forth, stopping frequently to sniff or paw at the ground. Though they move with easy, unhurried strides, their relaxed cadence propels them along at surprising speed. Soon the pair is within a hundred yards but still seems unaware of our presence.

"What should we do?" I whisper to Pat. The wolves are approaching from the side and are likely to see us if we don't find a different place to hide. We aren't hiding to escape but so that we can continue to watch them. Unlike an encounter with a bear in close proximity, we can revel in the presence of a wolf without the fear of being eaten. A free-ranging wolf is the embodiment of wildness, but only in the rarest of circumstances are wolves aggressive toward humans.

Pat points to a rock nearby. We slither on our bellies to this new hiding place and sit perfectly still, waiting for another sighting of the wolves. When several minutes pass and we still haven't seen them, Pat decides to poke his head out for a look. Immediately, he ducks back down.

"They're *right* there!" he whispers. "And I think they saw me." We peer from around the boulder and the wolves stare back at us, only fifty feet away. They look more curious than afraid, but they begin to retreat downhill, pausing every few strides to turn and glance at the two strange creatures huddled on the ground.

When the wolves are out of sight, we stand and stretch. I lean into Pat for a hug and he leans back. In the excitement of seeing the wolves, we have almost forgotten the morning's argument. By the time we reach the valley below, the light is fading to dusk and we find a place to camp along a small creek.

For the first time in months, darkness begins to punctuate the long Arctic days. I build a fire on the gravel bank and we huddle around the flames, smiling as we recount the curious stares of the yellow-eyed wolves. Later, lying together in the tent, we nest our bodies against each other. I begin to rehash the reasons behind our fight, wavering between an apology and an explanation, but think better of it and quit talking. Pat lets silence fill the air. Already, darkness has offered all of the forgiveness we might need. We have seen enough, together, in the past few months, to accept that we are each doing the best we can.

We spend our anniversary picking our way over a steep mountain pass, sidling carefully along a rocky ridge and across an ice-covered gully. The wind stirs up clouds of spindrift, and my feet turn numb as we kick steps in the snow up the final slope. When we reach the top of the pass, we're surprised to find a male gray-crowned rosy-finch chittering merrily, seemingly at home among the rocks and ice. It feels like we're being toasted at our own private anniversary party, with only a single bird in attendance. Compared to our dirty clothes and wind-chapped faces, his summer attire—a unique combination of brown, gray, and pink plumage—looks dashing. Unlike us, he also seems entirely unperturbed by the wind and blowing snow.

From glacial moraines to mountaintops, these chunky finches are often the only signs of life. Today is no exception. As we pause to watch, the bird flits and hops toward us, finally settling on a nearby snowfield. He burrows his head, then spreads his wings and nestles his rump in the snow. He shakes and preens, arranging each tail feather carefully. At the top of a steep pass, with several thousand feet below us on either side, this barren landscape hardly feels like a suitable place to raise a family, but that is probably exactly what he's been doing here. Gray-crowned rosy-finches nest at higher elevations and in more inhospitable terrain than just about any other species on the continent. Their alpine proclivities mean that relatively little is known about them; simply finding their nests requires robust mountaineering skills and a good set of lungs. Oddly, despite their remote summer residences, these birds seem to have almost no fear of humans, and our friendly companion repeatedly lands just ahead of us, playing a game of leapfrog as we continue to hike.

That evening, celebrating with tea and curried couscous, one of our favorite rotations among the six or eight meals that have made up our menu for the past several months, we drool over the luxuries that wait just a few dozen miles away. We are headed toward the first road we will cross in more than a thousand miles, the only one that penetrates the Brooks Range, otherwise stretching uninterrupted from the Yukon to Alaska's western border.

The next day, when we crest the final ridge above the in-famous North Slope Haul Road, a gravel-lined corridor built in the 1970s to support oil drilling on "the Slope" and for-mally named the James W. Dalton Highway, we stare out at a faint line of dust running through the mountains. Parallel-ing the road is the shiny hulk of the Trans-Alaska Pipeline, an eight-hundred-mile snake that bisects the state. Shuttling crude oil from the treeless Arctic coast to the temperate rain-forest of Prince William Sound, the pipeline is the artery of the Alaskan economy.

During construction and well into the first decade of oil ex-traction, the Slope was the cash cow that almost everyone drew a bit of milk from upon arrival in Alaska. Talk to most men (and many women) of my parents' generation—conservative or lib-eral, pro-development or pro-environment, wealthy or poor— and if they are longtime Alaskans, they will likely recall a stint on the Slope. My father was no exception. In 1974, freshly married and with a degree in civil engineering, he found him-self, along with hundreds of other young men looking for work, flying over the Brooks Range, one of the world's most intact ecosystems, to arrive at Prudhoe Bay, one of its largest oil fields.

Those of us born after this boom time never knew Alaska before her barrel prices became synonymous with economic stability. We were raised instead in a culture of the Alaska

Permanent Fund dividend—an annual payment to all Alaska residents that is an attempt to share the wealth generated by the sale of petroleum rights to multinational corporations. In short, we have all been bought out. I often wonder if the opportunity to have witnessed this landscape truly intact, even if only for a moment, might have been worth the heartbreak of watching as a pristine wilderness area was transformed into an industrial oil field.

Pat and I cross beneath the pipeline and angle onto the road, following the tracks of an Arctic fox until they disappear into the hard ground. Over the next hour, two trucks rumble by. One of the drivers waves; the other barely seems to notice the oddity of two hikers walking down a road that sees little traffic besides semis and the occasional tourist RV. We're six hundred road miles north of Anchorage, and more than one hundred miles north of the nearest town—if a place with an official population of twelve can be counted as a town. I crane my neck at the men as they whiz by at thirty miles an hour, gathering scattered details of their clothing and hair and beards. After so many weeks alone, ordinary sights have become spectacles. When another truck passes too close, kicking up small pebbles that rattle against our packs, we curse and cough in the cloud of dust and exhaust.

The road itself offers little reason to celebrate, but today our thoughts are occupied by more tangible matters. My dad had arranged for a resupply drop from our friend driving north to hunt caribou in the Brooks Range. Somewhere, just two or three miles away, a bear-proof barrel with food and equipment awaits us on the side of the road. For days we have been dreaming about food that has not been dehydrated or freeze-dried or stored in bags for months. Most goods in the sparsely stocked stores of remote Arctic communities boast higher concentrations of preservatives than anything resembling nutritional value. Peanut butter and jelly in a squeeze tube. Fritos. Twelve types of soda. Reese's peanut butter cups. White bread. Canned chili. Pizza sauce. If an item can't survive several trips by plane and boat and four-wheeler, it's simply not worth the effort of transporting. We haven't tasted a bite of fresh produce in nearly six weeks, and the only variety in our mundane backcountry repertoire has come from the treats my mom has occasionally smuggled into our resupply packages, already stale before they've reached us.

Another hour of hiking along the road brings us to the place where we hope to find our resupply. Our friend had left a detailed message on the satellite phone: *Just past a bend in the road there is a lone spruce tree perched high above a drainage. Behind the tree is your resupply barrel.* We pass the described bend in the road and I begin to salivate at the thought of what might await us. But at first glance, it seems there are a number of trees that fit this description. How will we know which one is correct? He provided GPS coordinates on the message, but I didn't write them down and our phone is nearly out of batteries. This is far from a crisis, but I'm not in the mood for a scavenger hunt. I want fresh food *now*.

I scan ahead with binoculars to be sure we haven't mistaken this bend for another. As I'm looking, Pat wanders to the edge of the road and peers across the tundra.

"It's here, I think," he exclaims. Sure enough, beneath a piece of orange flagging is the barrel. A note taped to its top reads, "FOOD RESUPPLY. Please do not disturb: we are relying on this to complete our trip. THANKS."

Pat unlatches the clasps and I slowly remove the lid. We take turns pulling off tape and opening the plastic bags inside, each step slow and careful, relishing in the ceremony of unveiling the barrel's contents. At the top there are notes from several friends, wishing us well on our travels and offering news from home. An anniversary card from my parents with a pair of sandhill cranes on the front reads, "Can't imagine two people better suited to walk across the Arctic together. This will certainly be an anniversary to remember!" One friend's hand-painted card features a miniature mountainscape with two tiny forms standing at the top of a snow-covered peak. Another friend has enclosed an assortment of tea bags with comments about love written on each one. Beneath the correspondence, besides our usual resupply items, we find two cans of beer, a sleeve of Pringles, two apples, a block of cheese, a loaf of bread, and a package of smoked salmon. A bag of hand-baked almond bars is wrapped in a red-and-white-checkered cloth.

As we examine our treats, a raven swoops in and lands on the top of a small spruce tree. It watches us intently, thrusting its head forward occasionally as though prodding us to unveil the rest of the barrel's contents. Because they're opportunistic by nature, ravens can make a home almost anywhere. Often thriving where humans are present, these birds consume discarded cheeseburgers, potato chips, dog food, and diapers with the same apparent satisfaction as berries, salmon, and moose carcasses. As a result, they're common visitors to fast-food parking lots and city dumps, but they can also be found at the remotest outposts—from North Slope oil fields to mountaineers' camps at seventeen thousand feet on the flanks of Denali. I once lost a sock to a curious raven at the base of a climb in the Alaska Range, where I thought Pat and I were the only two creatures present.

Ravens are not only scrappy, they're also smart, learning quickly to cue in on behavior and identify visual clues that might lead to food. If there is a free lunch to be had (or, in a pinch, a free sock), ravens will appear, seemingly materializing from the ether to join in the spoils. They likely developed these traits as a result of their scavenging nature; successful hunts by wolves and other predators occur sporadically, and meat is consumed quickly, so the birds must act fast when an opportunity presents itself. Today's bird clearly has a similar intention as it flaps to the ground, hops around the perimeter of the clearing, and pecks at the strap of my backpack where it rests a dozen feet away from us.

We lay out the cloth, pile our goodies on top of it, and stare at the feast before indulging. Toasting our first beers in more than two months, I feel the tingle of foam as it slides against my throat. Each bite of our meal is more delicious than the last—salty and sweet and crisp and fresh. It's the texture and contrast of flavors that I have missed most of all. When 90 percent of our food is some form of homogenized mush, anything that offers a bit of crunch might have come from a five-star restaurant. Right now, I can understand the cultural obsession with fine food and culinary arts. But I will never have a meal to rival this one—ever.

After we finish our lunch, I close my eyes in the sun, reveling in the fullness of my belly, the pleasant fatigue that comes from a day of hiking, the slight buzz from a single beer. I glance at Pat, sprawled next to me—shirt stained, hair coated with dust and leaves, a bit of salmon stuck to his cheek—and roll over to kiss his forehead. Four years of marriage. Three thousand miles. My body is stronger than it's ever been, my senses tuned to air and wind and water. Reaching Kotzebue before winter seems possible now, even likely. Since the scare of the Chandalar River and the frustrating days that followed, we've begun to move fast and fluid across the mountains. As we've walked, side by side, my pack slightly lighter than Pat's, I've felt every bit his partner. I lie back against the tundra and notice the raven above me again, this time holding an apple core triumphantly in its beak.

PINGALUK

We leave the Haul Road behind and enter a landscape that opens itself to us. Valleys dance beneath the autumn sunlight, and the ground holds firm and solid underfoot. The tussocks that plagued us for weeks have finally given way to bare, lichen-covered slopes. My muscles lengthen and flex with each stride. I have begun to understand what it means to live in constant motion. Like caribou. Like water. Like geese on their long annual migrations.

Anxieties that have surfaced along the way—of what will come after our journey is over or whether we will be ready to transition back to our "regular" lives—have faded again. As we walk, I think not about the fact of my dad's Parkinson's or my ambivalence about research, but about how much is possible. There are still a dozen ridges, four rivers, and three resupplies between us and Kotzebue. Even these facts feel less like a burden and more like a gift. The idea of worrying about only as much as I need to for today, or tomorrow, has finally taken hold.

Before leaving Bellingham, such narrow focus seemed impossible, even foolish. How would we possibly succeed if we didn't have everything carefully planned, our route detailed, the logistics ironed out? But uncertainty has become the only constant. Rather than adhering to schedules or itineraries, our days are shaped by the landscape; each valley, each river, each pass slightly different from the last. We can't know the weather, or the height of the bushes in the next valley, or when we might cross paths with a bear or a wolf. We are here, now, and that is enough. After almost five months, movement brings stillness. Movement brings peace.

Then comes the rain. It starts gently one afternoon, swaddling the hillsides in mist before giving way to a steady drizzle. Instead of views that stretch a dozen miles to the next ridge, at the crest of each pass we see only grayness. By evening the creeks have risen and our shoes squish against the soggy tundra. When we stop to camp, we roll ourselves into a dripping stand of alder, water seeping from the ground like a spring. Desperate to coerce wet wood into flame, we beat our paddle blades wildly at pyramids of waterlogged sticks. When the sparks refuse to catch, we eat granola bars and cheese under a bush, shivering as we imagine hot pizza and steaming soup.

On the third day of the rainstorm, water courses from our packs as we walk into the two-hundred-person village of Anaktuvuk Pass. This is the last community we will pass through until Kotzebue, our final destination, still a month and nearly six hundred miles away. We gravitate toward the village grocery store and ask the teenage clerk if there is a school or community center where we can dry out. She tells us that a construction crew is staying in the school but points to a Conex storage trailer with a plywood addition tacked to its side. "That's the hotel, inside the A.C. store."

When we reach the store, we open the front door to a narrow, dark hallway. Like most of the buildings in remote

Alaskan villages, this one is low and windowless, guarding against the winter wind and drifting snow. Inside, an elderly couple leans against paneled walls talking with the man at the front desk about caribou hunting. They smile broadly when we walk in, and the older man asks the usual question: "Where'd you come from?"

When we tell him that we hiked from the Haul Road, he replies, "Real good. Hiking keeps you fit. Look at me, seventy years old and I can still touch my toes." He demonstrates by kicking his right leg high into the air, making contact with his outstretched arm. I've never seen anyone his age move with such speed, and I instinctively jump back when his boot comes flying past my head. His wife laughs loudly next to him, revealing a gap where her right incisor is missing.

"That's right, we stay healthy by being out on the land. Where you headed next?"

"Down the John River, then over to Takahula Lake," I reply.

She frowns in response and asks, "Got a rifle?"

"We have bear spray," I say sheepishly, knowing the scorn with which the capsicum-laced bear deterrent is regarded by many rural Alaskan residents.

I frequently carry a firearm for fieldwork, a decision dictated more by policy than preference, but on personal trips we've almost always relied on bear spray. Not only are the logistics of crossing borders with a gun sometimes prohibitive, but even the lightest options weigh more than a packraft. Plus, bear encounters are often avoidable, and statistically bear spray has a much better track record. Both humans and bears tend to fare better when guns are removed from the equation. But of course there are exceptions.

"Ooh, it's not good to travel without a gun. Our people are caribou hunting and there's been *lots* of bears down south of here. You better be really careful."

The man at the front desk interrupts the scolding to ask if we want to buy something.

"We wondered about the hotel," I answer.

"Yeah. I'll show you." We follow the man down the hallway to a door that opens onto a space no larger than a typical utility room with a single bed crammed against one wall. The paint is yellowed with old water streaks, and shreds of wallpaper dangle from around the doorframe.

"Bathroom's down the hall," he says as he gestures. At nearly two hundred dollars, this would be the most expensive hotel room we've ever paid for, but as soon as I walk into the blast of the forced-air heater, I'm sold, outrageous price or not.

"OK, we'll take it," I say, glancing over at Pat, who looks similarly smitten by the hot air.

When I wander to the bathroom, I find cracked linoleum adorning the floors and walls and a mirror so warped I hardly recognize myself in the reflection. Two toilets are perched on odd plywood pedestals, providing a stadium view of the bathroom. The seats themselves are barely a foot tall and my knees touch my chest when I sit down to pee. The shower is similarly microsized, meaning that washing our hair will require a series of power squats. For the next several hours, as the metal siding whistles in the wind, we lounge on the bed and sort food, stacking bags full of grains, granola bars, and trail mix on the narrow bed. Our gear and clothes are strewn around the room, suspended from every makeshift hook we can find. Stripped down to our underwear, still damp from a scrub in the sink, we heat pot after pot of water in the electric kettle, downing most of a box of hot-chocolate mix while we pore over maps of the upcoming route.

In the morning, with the rain blowing sideways against the trailer wall, we slowly pack our bags, shuffling and reorganizing until we have exhausted all excuses to procrastinate. I hold my socks against the heater before pulling them on, letting the warmth soak into my toes. Through the window, the thermometer tacked to the outside of the trailer reads thirty-eight degrees Fahrenheit. Finally, at a quarter till noon, we drag ourselves outside and begin to hike south on a four-wheeler trail, each rutted track running like a stream in the deluge. Ankle-deep mud saturates our shoes and pants, which we had so carefully dried by the radiator overnight. Soon, the first trickle of water finds its way beneath my hood and down the front of my shirt.

By late afternoon we reach the confluence of Wolf Creek, where we'll leave the John River and head toward the crest of the range. We begin to hike upstream, but this drainage is flooded like all the others, banks crumbling, trees dangling by their roots from the shoreline. We have no choice but to ford the creek again and again. As we link arms and angle into the current, water creeps up our calves before reaching our thighs, then we feel the deep ache where ice meets groin. Each time, feet like stones themselves, I stumble along the cobbles of the bottom. The pain registers only later, once the numbness fades. When we begin to hike across the tundra again, a searing heat pulses in violent waves from my toes to my waist.

Pat and I find little to talk about, consumed by the task of getting through each painstaking hour. But even in our despair we're gentle with each other, offering touch as an antidote to what otherwise feels like a world made only of cold water and sharp edges. When I trip in a tangle of alders and fall onto my back, Pat runs to me, drops his pack, and lifts me from under my arms, as he might a child. He holds me tightly for several moments, silent, saying all he needs to with his body. *I'm here. We will be OK.*

For each of the next five days, the rain continues. Hard. Harder. The sky has mistaken itself for a sea. The only birds we see are small sparrows and thrushes, hunkered in the bushes waiting for a break in the weather to leave. It's raining still as we make our way across a brushy slope in the gathering darkness of the Pingaluk Valley. When a grizzly sow with three cubs appears on a ridge above us, I barely have the energy to care. She turns away from us, her cubs in tow, and I'm glad to see her go.

This is not how I had envisioned the western Brooks Range. In photos, the light is crisp and perfect, the sky so transparent it can hardly be called blue. All we have seen so far are oppressive black clouds that squash the grandeur. We're in a remote part of Gates of the Arctic National Park and Preserve, more than seventy miles above the Arctic Circle in a region that is technically defined as a "polar desert." Barely ten inches of precipitation fall each year, but apparently all of it is coming now. Even in mid-August, the Arctic equivalent of the cusp of winter, the day stretches nearly twenty hours long and night comes only as a shadow. But for the last week, it feels like the sun hasn't risen at all.

When we reach the confluence of two creeks, we begin to search for a place to camp. In the dark, noisy ravine, I feel trapped. Pat seems edgy, too, startling at the sound of my voice.

"There aren't a lot of great spots for the tent," I say with a pessimism that has formed in the rain.

"Yeah, but maybe this will work." Pat points to a mossy knoll that appears to be above flood range. We climb up to the site and find a fresh pile of bear scat as wide as a dinner plate, packed full of berries and grass. Next to this, obvious bowlegged footprints have been worn into the moss. This wouldn't be the first time we were forced to camp on a bear trail—like us, bears are drawn to the path of least resistance—but tonight I refuse. Maybe it's the rain, or the fact that we have already seen five grizzlies since breakfast.

"I don't want to camp here. It feels creepy," I say before

scrambling back down the way we came without waiting for Pat's reply.

Somewhere in the canyon, I hear the warbling notes of a northern shrike. Normally, I'm happy to encounter one of these uncommon birds, with their striking black masks and delicately hooked beaks. Tonight, two words echo in my mind: butcher bird. Similar in size to a robin, the northern shrike is a tenacious, predatory passerine that lures in smaller birds with song. *Right now,* I think to myself, *a bird is being called to its death.* Feeling unusually squeamish, I hope not to see the bird's larder. Shrikes employ a particularly gruesome form of caching, impaling or hanging dead animals from spines or branches. It's decidedly less charming than a chickadee hiding seeds under birch bark for later retrieval. Though their habits are necessary in a land of relative scarcity, and their methods of killing are more humane than most (they break their prey's neck, usually with a quick blow to the cervical spine), there's something eminently unsettling about seeing a dead warbler hanging by its neck from a spruce branch or a vole dangling from the crook of an alder bush.

We return to our packs and stand dumbly for a minute, rain dripping steadily from our hoods onto our noses. Unless we find a place to pitch our tent, we'll have to continue in the blackness. Pinned between a rock wall on one side and a steep tangle of brush on the other, this hardly seems like a tenable nighttime option.

"What about over there?" I point. A few patches of gravel remain above the waterline of the swollen creek. We choose a mound that sits slightly higher than the others, stamp out a mostly level site with our feet, and hope that the water won't rise overnight.

Once we burrow beneath the soggy down of our sleeping bags, I try my parents on the satellite phone, dialing twice before the call goes through. In the broken connection, I hear my dad's voice reporting "record rainfall," and "flooding at Red Dog Mine." Then silence; the call is lost. I dial again, craving the calm that I know he will offer.

"You guys are doing great," he says with the enthusiasm of a cross-country coach. We swap stories about the rain; mine from our dark, wet world, his from newspaper and television reports. He says that the forecast promises better weather to-morrow. If he's worried about us, he hides it well. He tells me that he's arranged the logistics for our upcoming resupplies and asks if we need anything else. His voice, steady and confident, reminds me that, like everything else, this too will pass.

Before I hang up the phone, my dad tries once more to lift the heaviness that is pouring from my voice despite my best attempts to sound cheerful.

"Things can only get better from here," he promises.

"And how are you?" I ask, but the connection fails before I hear his answer. I already know what he would say. *We're all doing fine here, honey. Just enjoying your adventure from a dis-tance.* If his back hurts or his tremor has worsened, he won't mention it. He has never been one for complaints.

I stare at the phone for several minutes, willing the black bars that indicate reception to return. But each call I try fails, and with no battery life to spare, I return the phone to its waterproof case without saying goodbye. We need to save the phone for emergencies. Zipped together in our damp sleeping bags, I reach for Pat in the dark. Though we haven't showered in more than a week, my sudden, aching desire for comfort is larger than body odor and sticky skin.

It's still raining when we wake up in the morning but the clouds feel brighter overhead. We find a small pile of dry sticks tucked under a boulder, light a fire, and heat water for coffee and oatmeal. When we've finished eating, we pack up camp quickly and head back up the hillside that will lead to the Pingaluk River. An hour into the day's hike, the rain eases and I drop my hood. As we work down the back side of the ridge toward the river, the bushes begin to thin and we can move quickly again, legs swinging freely downhill. We chat in intermittent spurts, mostly about the two things that occupy our thoughts—weather and food.

When we drift into a comfortable silence, we take turns hollering, "Hey, bear," or whistling to alert any furry four-legged travelers of our presence. Pat is leading this morning, choosing our route downslope toward the Pingaluk River. I follow just a few paces behind him to avoid the slap of bushes as they spring back into place. Soon, the rain eases completely and small patches of blue emerge from the western sky. I begin to hope that we might make it by evening to our next resupply point, at Takahula Lake, nearly thirty miles away.

As I'm wondering about whether we'll be able to paddle our boats on the river—*too much water? too little?*—I hear a faint rustling in the bushes behind me. It sounds large enough to be a jay, and I smile at the thought of an avian visitor. But then the noise gets louder and I hear the crackling of breaking branches.

I turn and face the sound. Fifteen feet uphill I see deep-set eyes, a pointed nose, and cinnamon-colored fur. The reality hits me with more surprise than fear. *Bear. Running. Toward me.*

"HEEYYYYY!" I have only enough time to shout a warning and instinctively throw one arm into the air, the other reaching for my bear spray, holstered on the hip belt of my backpack. This is not how I would typically respond to a bear in close quarters, where a calm voice and slow, nonthreatening motions are often the best way to defuse an edgy situation. But I realize immediately that this isn't a typical bear. We didn't cross paths accidentally, mutually startled into fight-or-flight. It's not a sow with cubs who feels threatened by my presence. This bear approached from behind, with a clear view of its surroundings. I know, almost before I've registered the fact that I'm staring into the eyes of a bear, that we're being stalked.

At the sound of my voice, the bear pauses. It stops less than a body's length away and stares directly at me. I feel for the plastic safety tab on the top of the bear spray canister and pop it off with my thumb, then aim. In the few seconds that have passed since I yelled, Pat has seen what's happening, grabbed his bear spray, and stepped by my side. We both stand momentarily frozen, canisters pointed, hoping that the bear will change its mind and bolt away into the bushes. But instead, it turns broadside and begins, ever so slowly, to walk past us downhill.

We move forward just enough to keep the bear in our sights as it ambles along, its casual gait belying the tension of the situation. Soon, it begins to circle back toward us. Again, the piercing gaze and unhurried steps. No signs of stress. No apparent fear. Only a singular, sinister focus. Us. I look around for anything we might use to defend ourselves, but there are no rocks nearby and the few sticks we see are half rotted into the moss-covered ground. We're left with two trekking poles, two canisters of bear spray, and our wits.

This is only the second black bear we've seen since crossing the Arctic Circle. Though it's large enough to be a small grizzly, this bear is missing the distinctive hump behind its shoulders. Its face is pointed, not dish-shaped, and its reddish-brown coat is one of many color variations common among both species. The fact that we're facing down a black bear, rather than a grizzly, does little to ease my nerves. Its four hundred plus pounds could easily knock me over with a swipe; its jaws could crunch through a limb as if it were a sausage.

Predatory bear attacks are exceedingly rare. But among the handful of such incidents that occur each year, black bears are often to blame. Unlike attacks by grizzlies, which are usually defensive in nature, maulings by black bears may be much less common but are almost always more calculated. A study of fatal attacks by black bears in North America found that 90 percent of such encounters were the result of predatory behavior by a lone male bear. Exactly like the one we're facing now.

Perhaps a lifetime of bullying by grizzlies has taught this bear that aggression is the only means of feeding oneself. Or maybe it simply sees the same reality that I see right now—it's clearly bigger and stronger than we are. We don't look much different in size from a caribou or moose calf. In this remote part of the Brooks Range, it's entirely possible that this bear has never seen a human before.

Now that we understand the situation, that the bear wants to eat us, our objective is clear. Somehow we have to change the bear's perception of who is the aggressor. As it approaches, we yell and raise our trekking poles in defense. Pat nods at me to get ready, and on the count of three we throw the poles at the bear. One hits it squarely in the jaw. It doesn't flinch. Rather, it continues walking toward us, eyes now focused on Pat. When its front paws come close enough to poke with our poles if we still had them, Pat fires. A stream of capsicum-pepper-laced spray explodes toward the bear's face. The bear sees the spray coming and ducks, contacting only the edge of the cloud. It sneezes and rolls its tongue like a dog spitting out a pill. But it does not run. Instead, it saunters away to the edge of the small clearing, and we spin around to watch. We have become matadors, minus the red flag and the crowd.

Suddenly, I want a gun.

By the time the bear comes for us the third time, the shock has worn off. Seconds stretch to minutes.

"It seems like if the bear really wanted to, it would get us," Pat says, panting.

"But this is classic predatory behavior…" I counter. Our voices sound small and distant.

"Like it's waiting for an easy opportunity?" Pat finishes my sentence. Yes, exactly. And we must still seem easy.

We yell again, and Pat fakes a charge, taking a few steps forward in an attempt to startle the bear. I feel spittle forming at the corners of my mouth as I rage against the animal that is intent on devouring us. This time, it approaches within twenty feet and pauses before skulking to the edge of the clearing. It sniffs the ground, glances toward us, and then looks away, as though we were merely an afterthought.

"Dammit," I say. "I can't tell if this is working at all."

Pat is silent for a moment, then replies, "Let's just watch for a minute more and see what happens. I think he might be getting tired of the game." We spin in a circle, following the bear with our eyes as it takes a full lap around the clearing's perimeter. After several minutes pass and it doesn't come any closer, I begin to hope that maybe Pat's right. The bear disappears into a stand of aspen and we scan the ground for the trekking poles we threw, picking them up one at a time, eyes darting nervously around the clearing. When I catch a glimpse of brown fur in my periphery again, I stifle a wave of nausea.

"The bastard is back." I point. We stand and watch, each poised with pepper spray in one hand, flimsy trekking pole in the other, as the bear shuffles away again. It comes into view a few minutes later in another clearing upslope. Not exactly a safe distance, but this is the greatest separation we've had from the bear since I first spotted it charging me. "Make a break for the river?" I ask.

Pat nods. "Probably our best chance. Once we get there we can either get in our boats or cross. Hopefully it'll lose our scent either way."

The five-hundred-yard walk to the river lasts an eternity. Pat leads us out of the clearing as I assume the rear position as bear scout. We hike in silence, stepping lightly, but our footsteps crack like artillery in the still air. Each cell in my body is alert for clues that we are being followed. Pat's head swivels back and forth as he searches for the quickest route to the river. I scan the trees behind us for a patch of cinnamon-colored fur, glancing forward only often enough to avoid colliding with overhanging branches. I strain to pick out sounds over the running water of the river. This is much scarier than facing down the bear. I'm discovering what it means to be hunted.

Finally we arrive at the edge of the Pingaluk River. As we link arms to wade across the thigh-deep current, I notice that the small video camera attached to Pat's backpack strap is flashing. Right after we spotted the bear, he had thought to turn it on. It feels eerily like a "just in case" last testimony, although I suspect that the filming was actually driven more by habit—whenever something interesting happens, Pat is quick to flip the video switch. He catches my glance and looks down, turning off the camera.

"Hopefully the excitement's over," he says, his voice flat. The water is cold and clear as it washes over our feet and calves.

When we reach the other side of the river, I want to collapse on the gravel. The water offers only a superficial division between us and our adversary, but at least it's something. I begin to think about the bear in the past tense. *There WAS a bear. There is not a bear anymore. It WANTED to eat us, but it didn't,* I tell myself.

The intensity of the encounter starts to fade as we hike downstream. What replaces it is the sickening feeling that I will never feel comfortable in bear country again. The dread of spending tonight and every other night in a tent. The thought of what would have happened if that bear had found us asleep in the dark ravine. After 156 days in the wilderness, I'm suddenly afraid of bears.

As though reading my mind, Pat says, "There's almost no chance we'll ever meet another bear like that again. Just think how many bears we've seen so far, and none of them have given us any trouble."

Logically, I know he's right. The thousands of days I've spent in bear country unharmed offer testament to the fact that bears are remarkably tolerant. On this trip alone we've counted forty-seven of them. And, until now, not one has behaved aggressively toward us. Surprised? Yes. Like the sow who startled from snacking on soapberries a dozen feet from us, clacked her jaw, then went crashing through the bushes out of sight. Others are curious, perhaps encountering a human for the first time. A small grizzly we spotted on the Arctic coast paralleled us for half a mile, standing on its hind legs repeatedly to sniff the air and peer at the odd, hunchbacked shapes on the beach. The bear on the Mackenzie Delta that ran toward us seemed merely confused. Given the opportunity, every bear we've met over the past several months has bowed out of a potential encounter before it escalated. Except today's bear. And it takes only one.

When the water in the Pingaluk River finally looks deep enough for boating, we decide to take our chances with the current in hopes of escaping the valley before dark. Each time I get hung up on a shallow bar or hop out to scout around a bend, I search the bushes for a bear that might come splashing in after us. All I see are shadows. Three hours and many boulders later, we reach the confluence of the Alatna River. It's wide and muddy and flowing faster than I can run. Paddling in the middle of the deep channel, basking in the evening sunshine, I finally relax. There is no bear that can reach me here.

TAKAHULA LAKE

Our next resupply box was sent in weeks ago with a man who is a caretaker at one of the only remaining plots of private land in Gates of the Arctic National Park and Preserve. We know nothing about him except that his name is Francesco and he's from Italy. And that he was kind enough to help us. Through arrangements with a local pilot, he agreed to shuttle a box of our food with his other supplies to his cabin on Takahula Lake. From the lake, we'll cross over the peaks of the Arrigetch Mountains to reach the Noatak River. The Noatak will take us to Kotzebue. It's the third week of August and we're closer to reaching our goal than I ever imagined.

But nothing seems certain anymore.

Before the rain and the bear, I had started to picture what it might feel like to land in Kotzebue. To paddle those last few miles. To go home. I was heady with confidence, strong and sure. Now I realize a line on a map is only that. We've planned our route around elevation contours and river bends, but we have no idea what we will really find. Everything can change in a day. In an instant.

The hike from the river to the lake is quick, less than a mile through an open forest of aspen trees, their leaves shimmering in the evening sunlight. When we reach the lake, which is larger than we'd expected, the location of the cabin is not immediately obvious. I stare into the clear blue water as we hop from boulder to boulder along the shoreline. After an hour of hiking, we spot the cabin across the lake and realize that we've headed in the wrong direction. At almost 10 p.m., I worry that it will be too late to disturb Francesco, who has had no forewarning about the specific date of our arrival. We inflate our boats and paddle toward the cabin, talking softly as we approach. But as we pull to shore, we hear the sound of hammering and see a barrel-chested man pounding sheets of plywood over the windows of the squat cabin.

"Hello," I holler from the water. He doesn't notice me at first, then looks up, startled at the sound of a voice. As soon as he spots us, he walks toward the lake with a giant grin and helps me pull my pack off my boat.

"Welcome, my friends," he says in a thick, resonant accent. "You must be *Car-o-leen* and *Pat-reek*. I am just working a little late because I have to leave tomorrow, sadly. But come in to rest. You are hungry and tired, I think." Familiar with the stress that accompanies an end-of-season departure from a remote cabin—packing, winterizing, securing the place against animals and weather—we realize immediately that we've arrived at a terrible time.

"Oh, you don't need to stop what you're doing for us," Pat says. "Maybe there's something we can help with?"

"No, no, please come. This is the perfect time to visit." As I step onto the shore, he extends his hand before enveloping me in a giant hug. A moment later, Pat receives the same warm embrace.

Francesco shows us to a wall tent where he invites us to spend the night. We drop our packs and follow him to the cabin, formed of hand-hewn spruce logs, each barely eight inches in diameter. Everything else about the place has been similarly built in miniature. I have to duck my head to enter the front door. Inside, Francesco looks like an oversize child in a dollhouse. But everything is scaled perfectly to the interior dimensions, and there is exactly enough room for the three of us to sit comfortably on benches that flank the wooden table. Francesco excuses himself, then steps outside to undo the work he has just finished so that we might enjoy the view that is quickly fading into darkness.

As we sip hot chocolate around a kerosene lamp, Pat asks Francesco how he finds the Arrigetch, meaning how does he *like* the range. But as Francesco begins to answer I'm reminded of this quirk of American English that easily gets lost in translation.

"Well, it was a tragedy that brought me to Alaska the first time," he says. In 1992, he explains, two of his close friends, famous Italian climbers, were lost on the Cassin Ridge on Denali. He came to help search for them and, in the process, discovered Alaska.

"Finding the Arrigetch a few years later was a funny thing," he continues. He explains that on his first trip, he and a friend attempted to hike in from the Haul Road, but the journey took so long they never reached the peaks. The next year they returned by plane to Takahula Lake, and Francesco met the couple who lived in this cabin. Twenty years later, Takahula Lake is his second home. We learn, too, that Francesco is a doctor by profession, and a writer and a painter by passion.

"But this is enough about me, my friends. I want to hear everything about what you have seen on your amazing journey."

I begin by telling him about the day that now seems like it might have happened a year ago. The dark ravine, the bear, the fear that we might not feel at home here again.

"Those black bears are the ones that cause trouble," he says simply. "Sometimes I have to try again and again to make them leave. But eventually they always do."

It quickly becomes obvious that Francesco knows this part of the Brooks Range well, perhaps better than anyone. Nearly every place we mention within a hundred-mile radius is familiar to him, though his modest descriptions downplay the impressive nature of his many solo adventures. When he speaks about traversing valleys and scaling steep rock walls, he's reverent in ways that our recent encounters with the rain and the bear have made me forget. In the quietest of ways, he's reminding me why I'm here.

By 2 a.m. we all begin to nod off and he wishes us good night, again delivered with affectionate hugs. A million questions still linger on my tongue for this fascinating man who finds his way again and again to a magical piece of wilderness many thousands of miles from home. As we head toward the wall tent, Francesco calls after us.

"You must stay again tomorrow and rest. This is your home as long as you like. There isn't much left to do to close up and I will give you instructions." As owners of an off-the-grid cabin ourselves, we know that inviting strangers to shut down the place for winter is a rare, trusting gift that extends far beyond the normal bounds of hospitality.

By the time we wake the next morning, Francesco has finished packing and his gear is stacked neatly by the water, where a float plane is scheduled to pick him up. Bread, honey, and tea are waiting on the table for us. He has written his e-mail address and phone number on a scrap of paper and tells us that we must come to visit him in the Dolomites of Italy. When we say goodbye, I find myself blinking back tears as I hug a man I've known barely twelve hours.

We wave until the plane is out of sight and then return

to drinking tea in the cozy cabin. I ask Pat if he thinks we should watch the video footage from the bear encounter. The last thing I want to do is relive this experience. But it's like the childhood ghost behind the closet door; I need to look before I can move on.

As we huddle over the tiny screen perched on the scratched wooden table, it's the missing footage that scares me most. For much of the twenty-minute clip, we see nothing but blurred tree trunks, Pat's shoes as they shuffle against the moss-covered ground, my waist and thighs in the same familiar brown pants I wear every day. Each time the bear steps into the frame, there's a flurry of movement, the camera unsteady as Pat lunges forward or throws his pole. As in an action film, the sound amps uncomfortably from the speakers and I reach to turn down the volume.

"Leave it for a sec," Pat says. "I want to hear what we're saying."

"Get out of here! Get the *fuck* out of here!" a voice that I barely recognize as my own hollers from the camera.

"You bastard. Get away from us. I'll kill you," Pat threatens as the bear's chest and legs come into view. Between the shouting, our voices are oddly calm. By the time I've seen the video clip for the third time, the shock of the encounter has eroded into a sort of boredom, our strident shouts no longer terrifying but annoying. It was just a bear. It was just one bear out of a thousand. It was a bear we will never have to see again. We put the camera away and pull out crackers and peanut butter for lunch.

As we're sitting down to eat, an odd sound drifts up from the lake. "Pat, listen," I tell him as I strain to hear what sounds like a human voice.

"Hello?"

This time it's unmistakable. We jump up from the table to

246

run outside. At the bottom of the hill below the cabin there's a man who appears to be in his mid-thirties, with a backpack drooping awkwardly from his shoulders, one arm slung tightly with a fleece shirt against his chest. He looks sheepishly at his arm as he introduces himself and explains that he dislocated his shoulder while boating. He leaves his backpack by the cabin and we follow him back into the forest. Within a few minutes we meet his friend, who is sweating heavily under a pack that must weigh almost as much as he does, towering above his head with a duffel bag strapped to its top. After dropping the first load at the shoreline, the four of us hike back to the river to pick up the rest of their gear and food. Along the way, we learn that James, the one with the dislocated shoulder, recently finished his surgery residency and is slated to start a new job in Vermont in two weeks. Aric, his childhood friend, works for the National Park Service and lives in Skagway, a short ferry ride from our cabin in Haines. The pair originally planned to boat down the Alatna River to a village where they would catch a flight back to Fairbanks, but because of the accident they will be picked up by float plane on Takahula Lake later this afternoon.

The men come into the cabin and share lunch with us, their smoked salmon, cheese, and cookies much more appealing than the stale crackers and peanut butter we have to offer. After eating, we walk down to the lake, where they assemble the foldable canoe they'll need to shuttle their gear to the float plane pickup. The fabric stretches easily over the clip-together frame, which is quickly transformed into a sleek-looking boat. I notice the intensity with which Pat is watching the canoe's assembly, and it soon occurs to me why he is so interested. For our final leg of the journey, boating down the Noatak River, we're stuck with either packrafting or using an equally sluggish inflatable canoe that we borrowed from a friend and arranged to have dropped with our final re-supply. The upper river is shallow and slow, and the lower sections are mostly flat water where a headwind could halt any progress. After our experience on the Mackenzie Delta, we're dreading the prospect of spending more time in an inflatable boat.

Neither of us is bold enough to suggest the outrageous—that we borrow a nearly brand-new canoe for a several-hundred-mile paddling trip from someone we've just met. But, without prompting, Aric asks if we have a boat to use on the Noatak. When we explain the situation with our pack-rafts, he insists that we take his canoe and offers to drop it with the flight company that will deliver our food and gear from Bettles, where the two men are headed next. Their generosity leaves me stammering thanks as we say goodbye.

Since we left Bellingham, our trip has been marked by kindnesses from strangers. Each time someone offers a hot shower or a soft bed, a cold beer or a steaming cup of coffee, I'm reminded that this journey is as much about human connections as it is about wilderness. At a remote lake in the company of just three other people, we've been given gifts of a cabin, a canoe, food, and friendship.

Back in Francesco's tiny cabin that afternoon, I notice movement out of the corner of my eye, look out the window, and see a gray jay peering in at me. At the sight of this familiar bird, I'm suddenly transported to my first morning with Pat at his cabin so many years ago. There, another jay had taught me something about love. This bird looks quizzically at me as it raps lightly on the Plexiglas. Perhaps Francesco, alone for months like Pat had been at his cabin, befriended this jay. Or perhaps the bird simply saw an opportunity for snacks from unwitting new visitors. Whatever the occasion, it's clamoring for my attention. It flutters back and forth in front of the window, then knocks again. My bag of trail mix sitting on the table in front of me consists mostly of peanuts; I already picked out all of the choicer items over the past several days. Staring at it, I decide that it's a fair trade. I don't have to eat the stale nuts, and the jay gets a leg up on its winter food supply.

I step outside and toss a handful of peanuts onto the ground. The bird stares at me, raises its crest to signal that I'm still not entirely trustworthy, and carefully picks up three nuts with its beak. I watch it fly to a nearby spruce tree and stuff the food beneath a piece of peeling bark. After it's finished, it returns for more, this time shoving several nuts between two logs of the cabin. Each time it caches, it looks around to see who might be spying. Satisfied that no squirrels, chickadees, or other potential robbers have spotted its hiding place, it flies back to collect another load. Apparently I'm not considered a threat, or perhaps not smart enough for such thievery, as my presence doesn't solicit more than a glance. We continue this game for another dozen rounds until my peanut supply is depleted. When the jay finally realizes I have nothing more to offer, it flies back to the first cache in the spruce tree, retrieves all three peanuts, and lands several feet away from me to eat them. I take this as a corvid's version of "thank you," nod, and head back inside to finish my tea.

Suddenly, I feel the first real pull toward home. I realize now that staying in one place is not the same as being stuck. We've seen so much in the past five months, covering ten or twenty or forty miles at a time. But this isn't the only way of seeing. Here, it's the seasons, the animals, the shadows and sounds that change. In the series of paintings Francesco had shown us before leaving, every view of the lake looked slightly different; each cloud, each shade of green, each reflection on the water's surface colored by the mood of the day. It takes much more than a visitor's eyes to uncover such subtleties.

That evening, I call my parents on the satellite phone. The answering machine picks up. "Hi. Just wanted to check in and let you know we're doing well." I hadn't told them about the bear encounter yet, and hesitate as I consider how much to divulge. "We had a little bit of a scare yesterday with a black bear that seemed to be stalking us. Luckily the bear decided to leave after we gave it some aggressive encouragement." I explain the bear's strange behavior and our defense with bear spray and poles. Then I tell them about Francesco and his hospitality and the new friends and canoe that appeared from the woods.

An hour later, I call back. Ever since we reached Francesco's cabin, I've been thinking about the last abbreviated conversation with my dad. There's something else I want to tell him. When I've needed it most, he has reminded me that things will get easier ahead. And, once again, after the rain and the bear, he was right. When I reach the answering machine this time, I keep my message short. "Dad, I also wanted to say thank you. I really needed your encouragement. And now, like you promised, everything is looking up. Even the sun has come out again."

CARIBOU CROSSING

We're racing against the arrival of winter, and it soon becomes clear that we're losing. At the head of the Arrigetch Valley, two days after leaving Francesco's cabin, a snowstorm on a critical mountain pass stops us. Knee-deep in dense snow, we stare out at slopes that threaten to avalanche and send us careening down sheer rock faces. After our third aborted attempt to cross the pass, we decide that our only choice is to try to find a lower-elevation alternative. With freeze-up approaching, we have no time to spare; temperatures have begun to drop into the teens at night and barely climb high enough by afternoon to melt frost from the bushes. We skitter back down the way we came, and I taste the bitterness of failure bubbling in my throat.

The detour we've outlined on the map will add dozens of miles through thick brush and across flooded streams, stretching our already thin rations even thinner. As we make our way slowly around the mountains, we hike on the stumps that used to be our feet, shoes soaked through with slush and the frigid waters of ice-laden creeks. After several days, I quit complaining about the throbbing pain, quit questioning the string of decisions that led us here. Desperation has a way of shifting one's perception: turning back is no longer a possibility. We're too low on food, the distances are too great. No matter how ugly, what I see in front of me is the only way forward.

And we're so close. Once we arrive at the Noatak River, we'll have made it to the last leg of our journey. From there, it's four hundred river miles to Kotzebue.

Late one morning, we reach the final obstacle separating us from the Noatak Valley—a mountain of boulders that cling to the earth at impossible angles. Drizzle has left their surfaces shiny and slick. Everything in my rational mind is trying to talk my body out of scrambling up the pass that looms above us. We knew from other trip reports that the route we had intended to take would work; we know nothing about this alternative. But in a remote valley of the western Brooks Range, more than one hundred miles from the nearest community, we have exhausted nearly all of our options.

Pat and I space ourselves widely to avoid each other's fall zone. As we clamber gingerly over loose, refrigerator-sized rocks, I try to be both feather-light and brawny at once, nudging each rock before committing my entire body weight to a single handhold. Every move is a decision. Every step has consequences. I shudder when my foot skids, then look over at Pat to exchange a hopeful, pleading glance.

We continue up, focused on our feet and hands. Gaining elevation by dozens and then hundreds of feet, the valley begins to fall away below us. When we finally crest the top of the pass, we whoop and holler and pose for a victory photo. But our celebration is short-lived. Down the other side, the north-facing aspect looks much worse than what we've just climbed. Instead of rain-covered rocks, the slope holds several inches of snow atop a thin layer of ice, and we stare out at a sea of white talus. As we start to descend, I wield the flesh of my butt as a shield, attempting to concentrate the brunt of my falls on the body part least likely to break. Progress is agonizing and I feel my hips bruising with each slip. Pat's thumb begins to bleed when the weight of his chest crushes his hand against a boulder.

Four hours later, scratched and bruised, we reach the heather. By now, shadows are darkening against the peaks, and I have to strain my eyes to see the ground in front of me. We're still much higher than we should be; at this elevation we risk getting caught in a blizzard. But the treacherous descent has left me weak and shaky. We scratch out a campsite among the rocks, pitching our tent by headlamp. Lying in my sleeping bag, I close my eyes against the blackness and soon hear the characteristic quieting of rain turning to snow. As I drift into sleep, winter knocks on the tent walls.

After our night of shivering, the morning arrives like a gift. The tent begins to brighten, and I can make out Pat's form next to me, head buried inside his sleeping bag, body curled into a ball. I'm layered in every scrap of clothing I have— two pairs of long underwear, hiking pants, three shirts, fleece pullover, synthetic vest, down jacket, balaclava, hat, gloves, extra socks on my hands—and I feel naked. When I unzip the tent door, I find my shoes frozen solid, laces encased in ice. I reach for my socks hanging from a tent pole above our heads. They, too, have stiffened, and shards of frost fall onto the tent floor as I crush them in my hands. I pull the socks over my unresponsive feet, then pry open my wooden shoes.

When I step out of the tent into the frosty morning, several inches of fresh snow shimmer in the bright morning light. Streams of water and ice cascade down towering granite slabs. For the first time in a week, I feel the sun on my up-turned face. Below, I see a red-gold valley, painted in every shade of autumn. It's more than a colorful landscape portrait, more than beauty itself. Because this is not just any valley. It is *the* valley, *our* valley. The Noatak Valley.

I poke my head back inside to share the beautiful morning. Pat is brushing frost from our sleeping bags with his glove. He looks up when I unzip the fly.

"It's incredible. I can see the river," I tell him. He rises onto his knees for a kiss. Even after waking to frozen socks and a sodden tent, we both know what this means. We have almost reached our goal: Kotzebue, the destination we imagined so many months ago. All that is left is to pick up our final re-supply at the river, stocked with plenty of food and a canoe, and to paddle downstream. Despite snow and flooding and a bear that wanted to eat us, we have made it through the mountains.

The celebratory moment passes as a fast-moving cloud obscures the sun and I'm startled back into the tasks at hand. Eat, pack, move. I hunt for firewood under several of the large boulders that dot the hillside, then fumble with the lighter, my cold thumb struggling to flick the ridged metal flint. The damp sticks produce only enough heat for luke-warm water, and we slurp down soupy oatmeal and cold tea. As we break down camp, we shove our gear easily into our packs—we are wearing the extra layers of clothing we nor-mally carry, and all the food we have left fits into a single bag that weighs no more than a quart of milk. I'll be relieved when we get our resupply, hopefully this afternoon, or tomor-row at the latest.

We traverse the narrow edge of a stream that steepens to a waterfall in places, and I moan over my burning feet. But soon the sun reappears and warms us enough to shed our down jackets. For the rest of the morning, our moods alter-nate between weariness and elation. We pause for a snack break and use the satellite phone to call the bush pilot who will deliver our resupply. He directs us to a slough just a few hours away.

By midafternoon, we reach the prearranged drop-off location. With the assurance that the plane will be here by evening, we decide to cook an early meal as a reward for making it to the slough. When we open our food bag to take an inventory, we find one pasta dinner, six granola bars, three tablespoons of olive oil, two packets of instant noodles, and a handful of trail mix. Because of the recent delays, we have been restricting our rations, splitting each dinner in half, spacing our snack breaks by three or four hours instead of the usual two. The sun shines on us as we devour an entire serving of pasta, our bellies more full than they've been since we left Takahula Lake. For dessert, we finish the remaining trail mix as we sprawl on the tundra and relax for the first time in days.

As afternoon becomes evening I feel the earliest flickers of concern. I call the air service again to confirm that they're still on their way. The woman who answers tells me that the pilots had been on hold because of weather for the past week, so they had to spend the afternoon shuttling clients who were waiting at the lodge to be flown out.

"But you told us that you would be here today," I counter.

"Yeah, but others have been waiting, too," she says.

Realizing the level of miscommunication that is quickly developing, I try to explain. "We're not really in a position to wait like people at the lodge might be. We're almost out of food and were told that the pilot would be here by now. The weather is still perfect."

"Well, the pilots are done for the evening, but someone should be there first thing in the morning," she promises. That night, I try to ignore an unsettled feeling that keeps me awake despite the exhaustion of the previous days.

In the morning I hold my breath as I peer out of the tent fly. A pale blue sky washes me with relief. When I call, the pilot answers and tells me that he is still grounded by fog on the other side of the range, but it looks to be clearing. For hours, we sit on the tundra and watch the clouds build. Finally, the breeze chills us and we crawl into the tent to wait.

Pat works to stay optimistic, distracting me by recounting our recent tribulations with the dramatic, abbreviated tales that I have dubbed his "explorer series," postcards he has been writing to his grandparents along the way. The most recent card read, "Dear Gram and Gramps, We are leaving Anaktuvuk. It has been raining and snowing. We have 600 miles ahead of us and winter is coming fast to the Arctic. We'd better get moving. Love, Pat." He hadn't intended for these messages to read like the journal of a solo nineteenth-century explorer, but somehow each one has become more punctuated by brevity than the last. As we try to recall the dramatic text from the last several cards, our stress is replaced briefly by laughter. But when the first drops of rain hit the tent, all humor dissipates.

By afternoon, the sky has darkened to a steely shade of gray and rain showers blow through like flocks of angry birds. I call the pilot only to hear what I already know. No flights until the weather improves. I put the satellite phone back in its case, cursing this piece of technology that offers a superficial link to the rest of the world. By evening, water pools on the tent floor and our shoes float in puddles inside the vestibule. With remorse at last night's indulgence, we have allowed ourselves only one granola bar each today, and I feel light-headed as I scurry to pack up camp. Moving as fast as we can in the pounding rain, we shuttle our tent and gear to a higher site. We agree to split a package of noodles in the evening. Rather than attempt to make a fire in the storm, we soak the ramen in cold water until we can wait no longer. As we eat, we listen to the crunch of uncooked noodles between our teeth and the unfriendly staccato of rain.

The deluge continues all night and into the next day. We take turns reading out loud from the first one hundred pages of a book I have carried as contraband among my clothes. On most legs, we agreed to leave books behind to save weight, a compromise I accepted because I'm usually too tired to read at night anyway. But after months of being deprived of words, I rebelled and have toted this torn half of a book for hundreds of miles through its namesake. *Alaska's Brooks Range: The Ultimate Mountains* tells the story of how, in the 1970s, the Alaska Native Interest Lands Claim Act secured more than 100 million acres of Alaska for the public trust. The book recounts the actions of a few key individuals who changed the trajectory of the state forever. The result of these conservation efforts is a land that has retained its fierceness. For several hundred miles we have been traveling across two of the largest protected areas in the country; together, the Gates of the Arctic National Park and Preserve and Noatak National Preserve encompass more than 15 million acres.

Sometimes, a landscape is granted the right to be left alone. The biological value of Arctic ecosystems is clear: there are rivers filled with salmon, valleys packed with caribou, patches of tundra that host millions of migratory birds, and plants found nowhere else on earth. But beneath these attributes is a simple fact of the human psyche: we all need to know that, somewhere, it's still possible to lose ourselves in the wilderness. The Brooks Range is exactly that kind of place.

The book offers a short reprieve from the reality of starving in a rainstorm. Even now, with only a few hundred calories of food remaining and no way out of what could be a long and desperate wait, Pat's eyes shine as he hears tales about other adventures in these mountains. I'm inspired by the idea that, as a biologist, my efforts might contribute to preserving the places and wildlife I love. I take comfort in Pat lying next to me and try not to think about the reality that is just now setting in: The plane might not come in time.

Without food, we don't have the energy to cover the nearly 150 miles necessary to reach the nearest village. Another week and we'd barely be able to move. The skies have to clear sometime, but there's no telling when that might be. The cusp of winter in the Arctic is a terrible time to wait out the weather.

I step outside to pee and spot a grizzly grazing a meadow on the other side of the slough. It doesn't seem to notice me, and I slip quietly back into the tent, hoping the grass is enough to satisfy its appetite. When the rain eases we unzip the tent to find that the bear has moved on, and we decide to make our own attempts at foraging. We don rain gear and optimistically grab several large ziplock bags. Red-leafed blueberry bushes blanket the slope near camp, but most of the berries have already withered and fallen off their stems. All that remain are crowberries that stain my fingers purple like a thousand tiny bruises. I fill my mouth again and again with their bland flesh, mealy against my tongue, but they do little to satisfy my hunger.

A male willow ptarmigan is foraging nearby, his head and obvious red eyebrow bobbing as he pecks at berries on the ground. Contrary to the instincts of most other avian species, this bird has no intention of leaving. Though ptarmigan might not win any prizes for their smarts, there's little question about their toughness. Unlike us, these birds can survive on willow buds and crowberries, in ground blizzards and ice storms. With feathered feet that act as snowshoes and perfectly white plumage, ptarmigan are winter specialists. During the coldest months, they spend up to 80 percent of their time in burrows dug into the snow, shivering to generate heat.

Seeing the ptarmigan only makes me realize how ill-suited we are for this environment, how poorly prepared we are for winter. We have no feathers for insulation, no specialized gut for digesting shrubs, no snowshoes on our feet. As much as my heart's not in it, I can't ignore the fact that a ptarmigan would make a decent meal, which we need desperately. I look around for a rock but by the time I've found one, the bird has taken cover in the shrubs out of sight.

By the next afternoon, the effects of fasting set in. I feel dizzy when I stand. My breath comes in short spurts as I stumble across the tundra and my heart races when I walk back to the tent. The river has turned to a muddy torrent overflowing its banks, making fishing, already lean along this stretch of river, nearly impossible. Adjacent to camp, the small slough holds nothing more than lily pads and water striders. We don't see the ptarmigan again. We stick to our plan to split only one granola bar and a tablespoon of olive oil a day as the knowledge of three more bars and another package of instant noodles in a bag at the foot of the tent gnaws at our bellies. Pat suits up to search for berries, but we have cleaned out our nearby patches and it takes him nearly an hour to gather several handfuls. When he returns, he's soaked and has clearly burned more calories staying warm than the berries he's managed to collect will provide.

For the first time, I understand what a chickadee must experience each night the temperature drops well below freezing and its survival hinges on the slightest of reserves. To persist in temperatures of thirty or even fifty degrees below zero, these tiny birds burn through 10 percent of their body weight in fat simply to make it through another day. There is no time to waste during the short daylight hours of a subarctic winter; every moment is spent in pursuit of food. At night, they find cavities in trees and enter a state of torpor, which allows their metabolism to slow and their core temperature to drop, preserving only the most vital of functions. As humans we can't enter torpor, nor are we as familiar with the threshold between hunger and starvation. I know only that we need food, and soon.

By the end of the third day, we occupy a hazy space between waking and dreaming. Just climbing out of the tent feels like an expedition. I've never been so weak before or had my mind so clouded by hunger. I waver between periods of heart-thumping anxiety followed by the sort of acceptance that borders on defeat.

"What if the plane doesn't come?" I ask for the thirtieth time.

"It will," Pat says. "It has to."

And, later, it's his turn to wonder. "Maybe we should have kept going on the first day. At least we'd be downriver where someone could reach us by boat."

"Maybe, but how could we have known? We had so little food already and they promised they would be here," I tell him. We both know that we've passed the point of indecision and our only choice is to wait.

I'd considered the possibility of dying on this trip, but I'd never imagined it might be here, like this. After facing avalanche-prone slopes and a predatory bear, floundering in an icy river and flipping a raft in the Arctic Ocean, starvation seemed like the most improbable of ends. It had never crossed my mind that we might face a slow and plodding death, where terror comes not from blood and gore or bodies swept away in a cold current but from the endless act of waiting. If I had thought more carefully about the history of the Arctic, perhaps I would have come to a different conclusion. Among the many tales of explorers who perished before us, starvation features prominently. Only now do I remember the stories of travelers eating shoe leather and, in some cases, eating one another. I recall a misguided expedition to Wrangel Island in which Ada Blackjack, a plucky Inupiaq woman, ended up the lone survivor after watching others lose first their hair and teeth and then their minds. I think about the young man from *Into the Wild* who walked into the wilderness, took up residence in a bus, and never came out. It's easy to be critical from a distance, to assume that such mishaps are an avoidable result of bad judgment, but suddenly we have found ourselves in a similar situation.

As the hours stretch on in agony, my thoughts become irrational and panicked. Do we face a slow and painful decline into delirium? Does the aching ever stop? Who will succumb first? Perhaps an animal will come to eat us before starvation takes the final swipe. We lie in the tent for hours at a time, moving only when we must. There is nothing to distract us, and waiting makes my mind spin more desperately out of control. At night, darkness amplifies my fear. When the tent rustles in the breeze, I imagine a hungry bear stalking us. I see shadows that aren't there. Again and again, I think I hear a swan calling faintly, but it's always gone when I open my eyes. Finally, I wake Pat to ask him how long he thinks we can last here.

"At least another week, maybe two," he says. "It wouldn't be pleasant, but I think we would survive." My stomach begins to cramp and convulse and eventually I resort to swallowing a small dose of painkiller as a sleep aid.

The next morning, more drizzle, another call to the pilot, who tells us that the weather is still not good enough to fly, followed by another day of hateful waiting. We sleep as much as we can now, saving our energy, preserving our sanity.

Codeine we've carried with us for emergencies helps to quell the persistent pain in my stomach. We peer out of the tent every hour, but nothing changes.

Around noon, the tent's beige fabric brightens and I squeeze my eyes against false hope. The clouds have remained low and stationary all day and the forecast calls for more rain. To wish for anything else at this point feels foolish and I steel myself for another round of despair. But when a harsh yellow light probes my closed lids and refuses to go away, I sit up and unzip the tent fly. Pat is alert now, too, looking at my face expectantly. My cracked lips stretch into a smile. Pat smiles back.

Within minutes I've reached the pilot with a weather update. "PLEASE COME. NOW." I plead. "We can see all the way over the tops of the peaks, the sky is blue, the winds are light." *And we are HUNGRY, hungrier than you can possibly imagine,* I think but don't say. On the other end I hear hesitation. The sunshine has sparked desperation inside of me, and I push a little harder. "We've been waiting for almost five days now and we really need you to come. We're starving. Please give it a try."

"OK," he says, "I'll be there as soon as I can."

We listen for the drone of a distant plane, willing insects, birds, anything that moves, to be the single sound that will save us. But two hours pass and the sky remains silent. I call again. I learn that they have been running flights close to the lodge, waiting for the weather to improve before heading to the north side of the range.

"Listen, we *are* on the north side of the range and the weather is perfect," I insist. "I don't know what other information you need."

Finally, after a long pause, the pilot asks exactly where we are. I can't believe it. We've been sitting in the same place for days—the same place where I'd originally been directed and then subsequently described in detail on several occasions. He wants to know how deep the slough is, to confirm that it will be adequate for landing, an inquiry that is equally unnerving.

I resist shouting at him and answer his questions as carefully as I can. And then I ask, "So, are you coming now?"

"Yes, I'm on my way."

Maybe he means after lunch, or after some other errand, but it takes much longer than the hour and a half flight time before we hear the sound of a distant engine. Almost as soon as we hear it, the sound gets fainter and disappears entirely.

Pat and I stare at each other and begin to blurt out angry questions. "Did they turn back? Did they misunderstand where we are?" The sun is still shining, but a few clouds have begun to settle on the adjacent peaks and I feel as though I'm suffocating.

When we hear another low buzzing sound an hour later, I guard desperately against the crushing disappointment of the last encounter. Only when the plane comes into view and heads directly toward us do I jump up and begin to tremble. The sudden motion draws a blackness over my eyes, and I kneel quietly until I feel my pulse returning again. As the plane circles and lands on the slough, we wait to meet the pilot on the bank.

"You have no idea how happy we are to see you," I say. And he doesn't. He's courteous but nothing more. He hands the boxes out of the plane and we shuttle them along in a fireman's carry. I have to steady myself against a bush to avoid fainting from the exertion, but soon we have unloaded our two boxes of glorious food, extra clothes, camp stove and fuel, Aric's folding canoe, and paddles. We open the first box before the plane is taxiing and cram Snickers bars, peanut butter, crackers, and trail mix into our mouths in a frenzy. Within minutes, we've eaten more than the sum of the previous week combined. I stifle a gag, my stomach shocked by the sudden onslaught of food. By the time we unpack the boxes and cook dinner, it has started to drizzle again. I wake in the early-morning hours to a chorus of wolves. When I peer outside, a nearly full moon rises between the clouds, casting the snowy peaks in a silvery glow. Just before dawn, a skein of snow geese passes overhead, their voices joining chorus with the river. We're ready to keep moving.

In the morning we tug the loaded canoe through standing reeds to the riverbank. It's raining again. I glance back at our site for a final time before the river carries us away. With the recent storms, the snowline has crept farther down the hillsides, and the air temperature hovers just above freezing. We pull to the bank for rounds of jumping jacks and push-ups to warm ourselves. Previous river reports had warned of shallow water and slow current on the upper Noatak, but the flooding continues and we're flushed downstream with purpose. Even with fierce headwinds, we cover almost 150 river miles in three days. On the first day, we spotted five grizzlies from the canoe, but since then the banks have been empty. Weary from the rain and the nagging cold, I focus my attention on the murky water as it swirls beneath the canoe—hood up, head down, and paddle. Occasionally I remind myself to look around, but mostly there isn't much to see.

As we round a bend in the river, I notice what appears to be a branch floating downstream. And then another. By the time I realize what I'm seeing, two caribou have landed on the far shore. They prance and shimmy, water flying in beads off their coats. When Pat points to the bank nearby, my breath catches in my throat. Dozens of animals stand at the water's edge, poised to cross. They glance at us as we pass, but their eyes are focused on the opposite side of the river. Quickly, we search for an eddy and paddle to shore. Soon, we hear splashing. As we peer over the bank, the caribou high-step into the river. When the water reaches their chests, they begin to swim. Cows and their calves pair tightly in the swift water, floating head to tail as they exchange quiet, reassuring grunts. The bulls' racks protrude dramatically from the river's surface. Many have started to drop the fuzzy brown "velvet" that covers newly formed antlers, and the effect is startling, shiny white antlers dangling bloodied strips of flesh that trail in the water.

After the last of the animals has crossed, we haul the canoe higher up the bank and walk over to the flurry of tracks they have left in the sand. A narrow trail leads into the bushes, and we hike up to investigate. The air is thick with the musty smell of caribou where hair and scat plaster the ground. We begin to hazard guesses at the number of animals that we'd seen. Fifty? A hundred? More? And then, as we stare down at the tiny tracks of a calf, a wave of sound approaches like a squall across the water. Soon, we hear brush crackling, and the wave becomes a tsunami.

"Quick, get down," I say, and we duck into a stand of willows. Within seconds, we're surrounded by caribou. The tendons in their legs click audibly and their breaths come in snorts and throaty huffs. In single file they march in front of us, so close I'm tempted to graze their flanks with my fingertips. Instead, I close my eyes and feel the steam rising from their bodies. Mostly, the caribou ignore us. But as they pass, some glance sideways in surprise, the flashing whites of their eyes sharp against deep brown irises. One large bull steps gingerly over Pat's outstretched legs. A curious calf sniffs us, its face mere inches from ours. Limbs frozen in place, Pat and I communicate with raised eyebrows and whispers.

"This is the single most amazing thing I've ever seen," Pat mouths to me. When I smile at him, I can see the emotion pooling in his eyes.

It's suddenly clear to me that the hungry wait, the rain and snow, and the fact that the pilot took days to reach us happened so that we could witness this migration. For months we've traveled in the shadow of caribou; first the Porcupine herd, then the central and western Arctic herds. Their rutted tracks, extending like veins across the landscape, guided us over the foothills and peaks of the Brooks Range, across hundreds of miles of Arctic vastness. Again and again, they led us away from steep cliffs and treacherous slopes, through terrain that seemed impenetrable. We learned to trust their wisdom over our own, to follow hoof prints as they traced an ancient route across the tundra. Now we are here with them.

For all of its seeming cruelties and callousness, the land has given us what we need most. Closure. Completeness. We never could have guessed that this glorious moment would be the culmination of our hardships.

Photos of caribou migration taken from the air show tens of thousands of animals moving in parallel ribbonlike tendrils fanning across the tundra. What we see now is more jumbled, more chaotic, the embodiment of motion. As we sit hunched beneath the willow bushes, the herd's dynamics unfold at our feet. Each animal's actions are clearly driven by something larger than itself. Even at its most hectic, the crossing is far from a stampede—I don't see any animals pushed into the river or even jostled by the others. Still, the herd's purpose is absolute. The same cloak of winter that chased us through snowy mountain passes and slush-filled streams hovers over the caribou.

Some of the animals continue down the trail out of sight, presumably to cross where we had seen the earlier band gathered. Others use an entry several feet from us that requires leaping off a sheer six-foot bank into the current below. Each animal faces a moment of indecision as its gaze darts between the river, the shore, and the other caribou nearby. The animals stack up from behind, and there is pressure to jump. When a bottleneck forms, it's almost always a cow-calf pair that first takes the plunge. The determination of the mothers is palpable. They're the ones with the most to lose. But they're also the ones who can't afford to delay. For calves, separation equals death, and each young animal holds tightly to its mother's side. Bulls and lone cows often hang back, deferring risk to the gangly youngsters and their travel-weary mothers.

We stare, transfixed, as wave after wave of caribou enters the river. Halfway across the quarter-mile crossing, the adults touch down on a shallow gravel bar where they prance and shake the water from their coats. The calves' legs are too short to reach the bottom, and they continue paddling, wild-eyed as they drift away from their mothers, carried farther downstream by the current. I hold my breath as one youngster is dragged past the eddy on the opposite shore where the other animals have exited.

"Pat, look. I think that calf is being swept away." As I raise my binoculars to watch, I anticipate a struggle, expect to see flailing limbs as the calf loses traction in the water and sees its mother slip from sight. But the calf's focus on the shoreline is complete, and only the occasional twitching of the erect white tail shows any sign of panic. As the bank steepens, the calf's odds seem to be shrinking. This is what happens, I remind myself. Nature is not always pretty. The current is fierce, the water cold. The calf is small and vulnerable. Pat scoots closer to me on the gravel bank until our thighs are touching. We pass the binoculars back and forth, watching and waiting. The calf is within a dozen feet of shore, but the water along the cutbank is turbulent and swift. Each time its desperate paddle strokes begin to close in on the shore, the current takes charge again. Just before the calf disappears from view, it manages to work its way to the edge of a crumbling mud bluff. I hold the binoculars perfectly still and narrate to Pat.

"I can't tell exactly what it's doing. It might be trying to flop onto the bank!" The small animal thrusts its chest forward and pedals its hooves into the deep mud. It's on its side, water rushing past. "It's fallen over, I think. Oh, no, the current is so close." Then, in a single, impossible motion, the animal channels its energy into its core and, like a spring released, launches onto its feet. Prancing, it high-steps along the steep bluff. When it eventually darts up the embankment into the brush, I clap my hands, the mother in me washed with relief.

For hours, we sit motionless, embedded in the intricacies of the herd's migration. Each time the bushes quiet and it seems there can't possibly be more caribou, another wave arrives. We lose count after the first several hundred animals, but thousands have passed. Finally, our stomachs begin to protest in hunger and the sky darkens to dusk. On hands and knees we sneak back to the canoe. We decide to camp on an island opposite the bank that offers a good view of the crossing. As we pitch our tent and eat dinner, the caribou continue to stream by. When we can no longer make out the silhouettes of their antlers bobbing along the surface, we crawl into the tent. Around midnight, I lie in my sleeping bag and listen to splashing and quiet grunts. I wake once in the dark and hear them crossing still.

By morning, they're gone. At the river's edge, beige caribou hair swishes in dense, floating mats, offering the only evidence of yesterday's spectacle. We pack up the canoe, then return to the place where the caribou had gathered. I scan the slopes carefully with my binoculars, searching for more animals lingering in the hills. We stand quietly and listen for the thrumming of hooves or the sounds of bushes snapping. But we're alone. Pat and I pause on the bank for a moment, shoulders pressed together, warmth seeping through our rain gear, and stare at the river that will carry us to the Chukchi Sea. To Kotzebue. To the end. And the beginning. If our journey ended now, I would have no regrets.

NOATAK STORIES

Fifty miles from the nearest village, we've stumbled unexpectedly onto a supersized version of Arctic hospitality. We bend over steaming bowls of soup, warming our toes next to the woodstove inside Ricky Ashby's 140-square-foot home. The cabin is perched on a bend of the lower Noatak River, nestled against a stand of spindly black spruce trees. We landed only thirty minutes ago in our canoe and are already mopping up our second helpings of soup with thick slices of homemade bread. As we slurp the salty broth and sip purple Kool-Aid from plastic cups, Ricky begins to count on his fingers in Japanese. *"Ichi, ni, san, shi, go."* He pauses, thinking for a second before he continues on the other hand. *"Roku, shichi, hachi, ku, ju.* My uncle taught me this when I was a boy. You'll never guess how he learned." Mouth full, I raise my eyebrows and widen my eyes in response.

He looks intently at us as he asks, "Do you have time for a story?"

Already, Ricky has piqued my interest: his remote, solitary existence; decades of international travel; dismissal of motors; role as a Quaker pastor. Plus, the sky's faucet is turned on full blast outside, our wet clothes are drying in the sauna, and the flooded river has pushed us far ahead of schedule. I have no desire to leave the warm cabin anytime soon.

We met Ricky after a morning of zooming through swirling gray-brown water, the current racing bank to bank from recent weeks of flooding. In the two days since passing the caribou, we'd seen little besides dense fog and a single sleeping musk ox that lifted an eyelid as we floated by. The days remaining to Kotzebue are few, but each one stretches on in the cold September rain.

Rounding a bend in the river, our hoods cinched tightly against the weather, we had smelled smoke just in time to turn the canoe into an eddy. We both paddled instinctively toward shore, drawn to the woodstove and the first cabin we'd seen since leaving Takahula Lake nearly three weeks earlier. As we dragged the boat up next to a set of stone steps built into the bank, I hollered hello into the mist. A man shuffled down the walkway, waving as he pulled on his green rubber rain jacket.

"Hi," I said. "Sorry for disturbing you. We saw the smoke as we were passing."

"No, no bother at all. Come in out of the rain." His shirt was unbuttoned beneath his jacket. "I was just getting out of the sauna and...oh, it's already afternoon," he continued, glancing down at his watch.

He waved us forward, then shook our hands eagerly. "I didn't expect any more floaters this year. You're here late," he said. The Noatak River, designated as a National Wild and Scenic River, draws several dozen recreational boaters each year. But by September, usually the only people left on the river are subsistence hunters from nearby villages and an occasional sport hunter flown in by bush plane.

We exchanged quick introductions with Ricky as he

bustled us up the ramp. At his direction, we left our wet clothes in the sauna after stripping to long underwear, then scurried to the cabin, where Ricky ladled soup into plastic bowls while telling us about himself. He was born in the village of Noatak, a full day's travel downriver from his cabin, and comes from a long lineage of Inupiaq people who have called the Alaskan Arctic home for millennia. For much of the year, he lives in this small cabin, a decision he said was prompted by wanting to live on the land before he's too old to do it alone. Unlike most Native Alaskans, who rely on snowmobiles and motorboats to hunt and fish, Ricky chooses to use only the power of his own muscles. His reasons for shunning motors echoed our own. Moving at human pace solidifies his connection to the land. When he slows down, he told us, he can see and feel so much more. Plus, he likes the physical challenge.

His life hasn't always been this way. Over the past several decades, he has served as a pastor and a drug counselor, has worked for the federal government and not at all. He now splits his time between the river, the village of Noatak, and various international destinations. He's a study in contrasts, a rare person who sees little conflict in owning many identities at once. When he mentioned that he had traveled to Southeast Asia the previous winter, I asked him to repeat himself. "Cambodia. It's a wonderful country. Have you been there?"

I shook my head.

As Ricky recites numbers in Japanese, Pat and I sit by the woodstove in chairs made of willow branches. After so many weeks of talking only to each other, we're out of practice around strangers. I can think of little to say, while Pat is unusually chatty, gushing over how delicious the soup is. Ricky doesn't seem bothered by our awkwardness. When he asks what has brought us here, he clucks appreciatively at how far we've come, but he isn't interested in a tally of miles. Instead, he wants to know if we've seen caribou, if bears have given us any trouble, and how we like this country. *His* country, he clarifies. He fixes himself a cup of tea and offers us the same before he settles again into his chair.

He begins slowly. "My grandparents lived near Barrow, in the northwestern part of Alaska. They had dog teams and trapped for furs to make money. One year, they decided to go to Wrangel Island to see if they could find any foxes."

Wrangel Island sits above Siberia's Kamchatka Peninsula in the northern Chukchi Sea. The three hundred miles of water separating the island from the Alaskan mainland are covered by pack ice in winter. Ricky tells us that his grandparents packed food and hunting and trapping supplies, and bundled several of their older children into dogsleds. They then set out across a sea of ice where temperatures regularly dropped to minus sixty degrees Fahrenheit, winds howled into ground blizzards, and polar bears prowled. It's a trip of massive proportions, especially with young children. Ricky explained their motivations in terms we understood. "I'm sure they needed the furs, but traveling is in our blood."

The family arrived safely, but while they were there, Russian soldiers showed up to enforce their jurisdiction over the politically contested island. With a language barrier and no documents to prove their identity as Americans, Ricky's grandparents and their children were forced to return to Siberia with the soldiers and later sent to Moscow. Eventually, the trading company cards they used to exchange furs for money helped to verify their residence. But by this time, there was no turning back; they were far from the Arctic coast. They then began a three-year westward trek that took them to some of the world's largest cities. After Moscow, they traveled to London, Tokyo, New York, and across the United States to Seattle. A steamship carried them north to Anchorage before they were finally able to return home to northern Alaska.

It's a journey I can scarcely imagine. Ricky's grandparents, who could navigate across three hundred miles of sea ice without map, compass, or motor, had to find their bearings in a world where survival had nothing to do with living off of the land. Decades before the existence of long-distance phone connections, Internet, or reliable mail service, their families must have presumed them dead. The fact that they were stranded on the other side of the earth fit no version of reality that anyone from a remote Arctic village would have known. Before I can ask for more details about his grandparents' journey, Ricky is up again, hurrying to check if our clothes are dry.

As we pull on our rain gear in the sauna's tiny Arctic entry, Ricky tells us that the fastest trip he ever made to town was twenty-two hours. Hearing this, I wish we hadn't dallied quite so long in the cabin. We had guessed that it would take us less than twelve hours to cover the distance to Noatak in the canoe. But when he continues, I realize that he's not talking about traveling there by boat in the summertime. "I think minus thirty is the perfect temperature. Not too hot, not too cold." *Minus thirty?* Instead of a river trip, he's describing an all-night, fifty-four-mile trip by skis.

"That was the best pair of skis I ever made," he says. "They were *fast!*"

Our visit with Ricky has left me humbled by the many versions of adventure, some planned, like ours and Ricky's, others unplanned, like his grandparents'. No matter the distance or destination, we're all joined by the most basic of human desires—to see what's around the next bend. Today, our journey feels neither remarkable nor petty. It feels simply and completely our own.

After leaving Ricky's cabin we paddle in the rain until dusk, then pitch our tent on a gravel bar dotted with bear tracks. As Pat unrolls our sleeping mats and pads, I light the stove to heat water for dinner, indulging in our most recent luxury. Included in our resupply drop at the headwaters of the Noatak River was a camp stove that burns pressurized white gas. Prior to this, we couldn't justify the weight of the stove, nor could we reliably get fuel at our resupply points. Now, our meals are ready in less than half the time it would take to build a fire. Our eyes aren't rimmed red with smoke, tea no longer has the persistent flavor of charcoal, and when it's pouring, we don't have to resort to gnawing on granola bars and crackers for dinner. As I'm stirring boiling water into a bowl of instant mashed potatoes, it occurs to me how easily we are satisfied these days. Tonight, this bent, sooty stove ranks among my favorite objects on earth.

The rain eases and we relax on the gravel, watching murky water swirl against the island's bank, mesmerized by the pattern flickering in the reflection of our headlamps as we shovel food greedily into our mouths. After we finish dinner, my thoughts drift to what lies ahead. When the trip ends and we return home, what will we wish for? Will tea from a ceramic mug, heated on an electric stove, still feel luxurious? Will I remember how few possessions I have come to need? Can I reconcile our time in the wilderness with a more ordinary existence, working as a researcher and living in a city? Perhaps I don't have to. Meeting Ricky has helped me see that a full life, by definition, doesn't fit into tidy boxes.

In the morning, I hear two northern wheatears fly over our camp. They pass low enough for me to see them clearly with my binoculars, and I can make out their beige juvenile plumage. It's late in the season for them to be here still, their journey ahead impossibly long. A recent study tracked the movements of wheatears using tiny transmitters and discovered remarkable feats of migration: birds from Alaska flew nine thousand miles to Kenya; those from Canada crossed the Atlantic to spend the winter in Mauritania. For young birds, just a few weeks old, this departure also signals the ultimate coming-of-age passage. Not only must they soar on brand-new wings; they somehow must find their own way. Adults typically depart before the fledglings, leaving them to navigate alone across oceans and continents. If the birds I see make it, they'll leave the company of caribou and musk oxen to sit on the backs of elephants and zebras. In their relatively short lives, they will travel as much as they will stay put. For them, as for us, there is more than one path that leads to home.

As we eat our breakfast, I think about migration, seasons, and the pulses of the Arctic. Now that I've seen the itinerant wheatears, the sedentary ptarmigan, the driven caribou, the stoic musk oxen all making their way in this land of extremes, I want to understand exactly *how* they do it. I want to ensure that they keep doing these things far beyond my lifetime. There's so much left to learn. I tell Pat I feel ready to be a biologist again. "Good," he says. "But I honestly don't think you ever strayed very far. You don't know how *not* to be a biologist."

From camp, it's a short paddle to the village of Noatak. We arrive before noon and park our canoe next to a dented aluminum motorboat with empty Pepsi cans littering its floorboards. I ask a man working on an outboard engine where we might find a phone in town. Our satellite phone is low on batteries, and I want to let my family know that we're nearing Kotzebue.

"Bingo hall," he replies and points down the dirt road. The bingo hall, which doubles as a community center, consists of side-by-side Conex trailers with several Plexiglas windows, fronted by a wooden staircase. We open the door slowly and peer inside. Half of the village seems to be gathered here, drinking coffee or chatting in groups at long plastic tables. Several teenagers are playing pool in the corner. A line extends from the surprising centerpiece of the room: a Lotto station selling paper pull tabs. The players pass over fistfuls of cash in exchange for brightly colored sheets adorned with miniature pictures of fruit, pigs, dice, and footballs. While we pause at the door, a young man waves for us to come inside and points at the coffeepot sitting nearby.

We step into the room and drop our muddy jackets on a chair. At the counter, we pour coffee into Styrofoam cups, then spoon nondairy creamer into the steaming liquid. Choosing a seat at a table with a communal phone, I pick up the receiver and hear a guttural voice in the earpiece. Across the room, I see an old man talking on the shared line from another handset. Although cell towers and wireless Internet are beginning to find their way even to the most remote parts of the globe, many Alaskan villages still rely on landlines and satellite signals. When he finishes, I pull out the calling card I have carried with us for exactly this purpose and dial my parents' number. There's a lull between each jumbled ring, and I'm about to hang up and try again when I hear my dad pick up. He sounds like he's speaking through a hose underwater.

"Hello?"

I answer, "Hi, Dad. It's Caroline."

A moment later, I hear "Hello?" again, followed by an excited "HI, CAROLINE!"

The delay is awful, but we wait between sentences to let the words filter through.

"I can't talk long because I'm on a shared phone, but we made it to Noatak," I tell him.

"When should I book my ticket to Kotzebue?" he replies. I assume he's joking.

"Hopefully we'll be there in three days."

"OK, that's the ninth. I'll get there on the eighth." Now I start to wonder if he's not.

"You're serious about coming to Kotzebue?" I ask. Kotzebue is 549 air miles and an expensive flight from Anchorage.

"Of course, I wouldn't miss it. Your mom wants to come, too, but she's in Whitehorse for the weekend." I nudge Pat, but he's in the middle of a conversation at the table next to me.

"You really don't have to do that, Dad," I say. I can't hear his reply over the sound of a low voice saying, "Hello? Hello? Who's there?" into another phone handset somewhere else in the hall. I try to interrupt and explain the line is still being used, but instead I hear the bleep, bleep, bleep of someone trying to dial out.

When I hang up, Pat is talking to the same young man who waved us inside. They're sitting close to each other, their voices low and hushed. A few minutes later, I hear Pat say, "Wow, thanks for sharing that with me." His eyes are wide and his cheeks are flushed, like he's just seen something amazing. We pull on our jackets and say goodbye, shaking hands with the people we've met as we leave.

As we walk back toward the canoe, Pat relates his encounter with the boy, who introduced himself as Lonnie. He begins to tell me the details, pausing occasionally to let out his breath like Lonnie had.

Last February, Lonnie was traveling home from a nearby village by snowmobile with a friend. They were caught in a storm, and then separated when Lonnie's engine failed. After Lonnie tried and was unable to get it to start again, he decided to continue on foot. Soon, his footprints were drifted in and he realized he was lost; he didn't know the way back to his snowmobile or to his village. The only choice was to dig into the snow and hope that someone would find him.

A local search party was launched when the boys didn't return that night; two dozen snowmobiles and a plane scoured the area. By late the next afternoon, they had located Lonnie's friend, cold but otherwise fine, and Lonnie's abandoned backpack and snowmobile, but they saw no signs of him. By the end of the second day Lonnie had begun to wonder if he would ever be found. On the third day he started to hallucinate. He said he didn't believe in God, but that someone spoke to him while he was lying there. He became too weak to sit or stand and could only see patterns of light and dark through his cocoon of snow. When he heard the sound of engines, he didn't know if they were real. Starved, dehydrated, half frozen, he thought to raise his leg. It would have been too late as the snowmobiles had already passed, except for one lucky accident: the last rider had dropped a gas can. When he circled back to look for it, he spotted Lonnie's leg hovering above the surface of the snow. When they dug Lonnie out, his feet were frozen solid inside his boots and his cheeks had turned black with frostbite. At the hospital in Kotzebue, the doctors weren't sure he would survive; if he did, it was certain that his legs would be amputated below the knee.

When we met Lonnie at the bingo hall almost seven months later, I didn't notice his gait. He wore normal shoes and pants. He stood on his own legs. Somehow, he had not only survived his ordeal but managed to keep all of his limbs. He had told Pat that he couldn't believe how many people had searched for him and how they never gave up. He said he still wasn't sure if he believed in God, but he believed in something.

In the last days of our journey, Lonnie's story resonates. We haven't traveled so close to the threshold of death. We never needed a rescue. I haven't found God. But we've known the power of the wild, the all-consuming demands of rain and snow and wind, the callousness of mountains and rivers. We've been cared for by strangers. We've felt part of something much larger than ourselves.

In life, we're always closer to the edge than we like to admit, never guaranteed our next breath, never sure of what will follow this moment. We're human. We're vulnerable. With love comes the risk of loss. There are a million accidents waiting to happen, future illnesses too terrible to imagine, the potential for the ordinary to turn tragic. This is true in cities and towns as much as it is in the wilderness. But out here we face these facts more clearly, aware of the divide between today and tomorrow. And, for this reason, every day counts.

FINAL DAYS

Half a day's paddle from the village of Noatak, the river moves like liquefied mud, opaque and heavy with silt. According to the map, we're only a dozen miles from the sea, but the channel meanders back and forth and seems more intent on taking in the sights than actually getting anywhere. Instead of mileage, we track our progress in river bends, snaking around each oxbow, paddling west, then east, then west again.

We're following an isolated finger of forest that traces a green swath along the Noatak River where mature spruce trees line the bank. Nudging up against the Chukchi Sea, which hosts walruses and pack ice, the forest feels like an anomaly. If I looked only at the trees, I'd swear we were deep in the Alaskan interior. But then a spotted seal, as marine a creature as one can find, noses up behind our canoe, whiskered snout blowing steamy puffs of breath into the cold air.

The river's slowing forces us to do the same. There are no rapids to negotiate, no navigation required to follow the wide murky channel. This leaves us with nothing but our own thoughts. For the first time in weeks, the morning's sun rises crisp against a pale sky, and its gentle autumn warmth slowly filters through our soggy clothes. I can taste the nearness of our destination, alternately acrid and sweet, and with this, the realization that everything will soon change. Instead of following river bends and seal snouts, I will be back in a world where time is a commodity. A world where there are too few hours to watch water flow on its own course, too many distractions to notice how the moon tracks across the sky.

Barring high winds or rough seas, tomorrow we'll reach the village of Kotzebue, not just another community along the way but the last. The end. From Kotzebue we plan to board a jet and fly back to Anchorage. Back to lives we left six months and four thousand miles ago. Back to lives that can't possibly be the same. I recall the months I spent camped in a cubicle, working frantically at the computer to finish the last of my dissertation. My eyes were always red and scratchy. I had to set an alarm to remind myself to look away from the screen every hour. For a time, I lost my connection to the outdoors. Now, I've forgotten what it feels like to be inside.

Because we're nearing the end, even familiar sights have become precious. I take photographs of the most mundane scenes. A moss-covered log that drifts past in the gray water. Our green and orange dry bags lashed to the bottom of the canoe. The hole in my gloves just above the right thumb. The stain of spruce pitch on my pants that has taken the shape of a hawk.

A caribou trots along the shoreline above us, and I explode with a string of curses when I can't dig out the camera fast enough. "I missed it," I tell Pat.

"Missed what?" he asks. "You were looking right at it."

"No, I wanted to take a picture. It will probably be our last caribou." We put our paddles down and wait to see if it will reappear from the bushes. But the caribou is gone. Even before they're over, I'm trying to fix these moments in my memory, afraid that they will fade like the current once we reach the ocean.

I've made several attempts to talk through the impending reality of this transition with Pat, but conversation fizzles before it even really begins. Pat doesn't resist the topic, but, like me, he has little to say. Over the past six months, we've rarely run out of things to chat about, no more bored with each other than with the ever-changing landscape. Today is different. There are no sentences to be strung together, no thoughts to be volleyed naturally from one of us to the other. This isn't because we disagree but because neither of us has found the words to express what it means to finish what we started so many months ago. To complete a dream formed a decade ago.

I feel like we owe it to ourselves to acknowledge the major change that is now just a day or two away. And we owe it to ourselves to celebrate a goal that once seemed impossible. I want us to revel, even if just briefly, in the satisfaction that comes with success. But now that we've almost pulled it off, the accomplishment seems only tangential.

For six months, I've woken each morning to something new. The smell of algae-smeared rocks of Pacific coves. The taste of new shoots of alpine heather. A bear intent on eating me. Caribou trembling as they pass inches from my face. I've heard stories of survival and adventure—from Francesco, from Ricky, from Lonnie, from the others we've met along the way. I've felt cold and scared and hungry. I've felt alive and loved and safe. Suddenly, the possibilities are infinite. And terrifying. We could cross another continent under our own power. Or an ocean. Or set off and never return to any former notion of home. If we had failed or ended the trip feeling battered and bitter, it would be easier to leave this all behind.

Even so, I know we can't simply keep going forever. We

can't even keep going through winter, at least not without completely re-outfitting ourselves. And a lifetime spent chasing our own version of adventure is hedonistic in ways that gnaw at us both. I don't want to isolate myself from the world. I miss my parents, my sister, my brother, the nephew I know only from a distance. I feel a responsibility to help protect the places I love so much. We might even be ready for a child sometime soon. But does this mean I must let go of everything I've found here? Does it mean that this is really *the end*?

I'm torn between two worlds, two versions of truth. Adventure or family. Wilderness or home. Nature or science. I'm beginning to wonder whether such divisions exist only in my mind. I paddle through a jumble of thoughts that pass like drifting bits of cotton grass, each swirling once above my head before disappearing.

By evening, the forest has vanished, the last scraggly spruce giving way to willow shrubs and estuarine grasses. In the distance, I can see the ocean and a glimpse of white, an ice floe sparkling in the evening's setting sun.

"Do you see that?" I ask Pat. By September, the ice should be gone, next winter's pack ice not yet formed. But as the words take shape in my mouth, I see the ice levitate and rise into the sky, shattering into pieces. Swans. Thousands and thousands of tundra swans, with golden necks and wings on fire. Their heavy steps patter against the water as they take flight, a gathering of angels against a steel blue sky. The birds circle once, calling in hoarse voices. Moments later, they drop down toward the water again, landing on a shallow bar. We pull to the edge of the slough and step out of the canoe to watch as their shimmering bodies are stretched tall by bending Arctic light. Soon, they will launch again, heading south to a place where the ocean doesn't freeze solid in winter.

Unlike godwits and other ultramarathoners of the sky, swans aren't built for endurance. At twenty-five pounds, they need to eat frequently and in large quantities to maintain their weight. Instead of a single, nonstop migration, they hop-scotch along the coast as they travel the two thousand miles to California or Nevada, where they'll spend the winter. This shallow inlet where we're all gathered offers food and rest. They might pause here for a week or more.

For us, there is just one last crossing—three miles across Kotzebue Sound. It's a big one to make in a canoe. Here, the Chukchi Sea funnels into a narrow channel where waves build quickly when the wind blows. If we go now, we can probably cross before dark. If we wait until morning, we run the risk that the wind will be too strong and the chop too large. Suddenly I don't want to leave. I don't want to arrive. I want to stay put, waiting here with the swans. For once, we decide to stop before we have to.

We set up camp against a chorus of bugles and flapping wings. In the dusk, our beige tent is a spot of drabness in an otherwise electric world. Green grass. Blue ocean. White swans. We're camped eight miles from Kotzebue, which, from this distance, looks like little more than a pile of drift-wood. Later, nestled in my sleeping bag, the darkness reminds me of the transition already under way. Summer to winter; light to dark; north to south. The birds gathered, ready to go. Our sun compass, pointing toward home. Each time I feel myself nodding toward sleep, I open my eyes wide and listen. There's nothing to fear tonight: no bears, no avalanches, no storms. But I'm not ready for the last night to pass. Pat is restless, too, and we lie side by side on our backs, the only sounds our breath and the calling of the swans.

I've never been poised in saying goodbye, and tonight is no exception. Even without looking at me, Pat can tell when I begin to cry. He's tuned to the tiny sniffles, the way I swallow just a bit more frequently than usual. But he also knows that I need this time alone. Only when my tears finally drip onto the sleeping mat does he roll toward me for a hug. I nestle against him, feeling his heartbeat against my back. I'm a jumble of happy and sad, satisfied but full of longing, thankful yet greedy for more.

I've woken to caribou tracks and birdsong more mornings than I thought possible. I've traveled by my own muscles more miles than I ever imagined I could. All of this with a man who loves me, who will help me face whatever might come next. Bears or snowstorms. A child or an illness. A job that might be right, or might not. Even though I feel like I'll be leaving a part of myself on this riverbank, there's nothing about this night that I want to change. I don't need to understand. I don't need to know what it all means. Tonight, I let myself simply be.

The swans continue to call, though the flock is quieter now. *We are here,* they seem to say. *We are here and that's enough for now.* I unzip the tent fly to look outside and see neon green brushstrokes of the aurora skitter across the sky. When I eventually drift into sleep, it's the same heavy slumber I've come to know recently. One last night of simple gratitude.

When we wake the next morning, the world is luminous. Every surface—canoe, shoes, grass, driftwood, tent—is coated with a crystalline web of frost, glowing in the sunlight. The ocean is the color of slate. The swans are gone. I slither back into my sleeping bag and curl against Pat. When he gently unwinds my arm from his back and steps out of the tent, I lie still for a moment longer.

We pack up slowly, lingering over each task that has become so routine. I begin to stuff our sleeping bags unceremoniously into their compression sacks, then stop to press a fistful of down against my cheek. As I kneel to roll up my sleeping pad, Pat lifts the fly off of the tent. He steps back and takes a photograph of me silhouetted against the rising sun. It will be one of the last images we have from that day.

I pull on my rain pants, patched with duct tape, and zip myself into the stained orange jacket that has doubled as my pillow for the last six months. Outside, Pat's footprints are etched into the frozen grass. I walk carefully, ice shimmying down the stalks as I parallel his tracks. He's hunched over the stove, and a cloud of steam billows around his head as he carefully adds two packets of instant coffee to the water bottle we've shared each morning for the last 176 days.

"Aren't you glad I won?" I ask, teasing Pat about our long-ago debate over whether we should have hot drinks on the trip.

Pat looks up at me and smiles. In another place and time, this watery mix would be barely palatable. Out here it's what gets me out of bed each morning. Soon, we'll have gourmet coffee and as many hot drinks as we want. What we won't have is everything I see around me right now. I push thoughts of gleaming grocery aisles and pavement and cars out of my mind. For one final morning, it's just us and our steaming water bottle of weak coffee. Pat has improvised a seat out of driftwood near the stove. I join him and settle between his legs, leaning my back against his chest.

I don't know if we'll have a baby soon, or at all. I don't know how quickly my dad's illness will worsen. I'm still not sure if I've made the right choice in accepting the research job in Anchorage. But I do know that Pat and I will navigate this terrain together, just as we have since we left Bellingham.

Even on what might be our last day of the trip, we can't afford to dawdle for too long. A light breeze blows and a few

whitecaps are building on the ocean's surface. This is our signal to go. Being stuck for a week only half a day's paddle from
Kotzebue is not the ending we hope to write. So we pack up
our bags and load the boat, looking back one more time at our
campsite before climbing into the canoe.

What starts as a choppy ride soon becomes smooth. The
waves flatten with each paddle stroke, the sea granting us a
final moment of reflection. The crossing, which could have
been stressful, instead helps to calm our restless minds. We
find a beach on the other side where we heat water for tea
and call our families. We convince ourselves it's easier that
way—telling others about the end before really reaching it. I
leave a message for my dad on his cell phone.

"We're almost at Kotzebue. I can see the lights and buildings. It's a little hard to believe. Hopefully we'll find you in
town."

When I hang up, I wonder if he'll hear what I really mean.
*I'm elated. I'm terrified. I hope you're the first person we see
when we get there.*

We stall a little longer before climbing back into the canoe.
We paddle in silence for the last hour, letting the water do
our talking for us. Splash, drip, drip, drip. Splash, drip, drip.
Steady and synchronized, it's a rhythm we know well. When
we arrive at Kotzebue an hour later, I see my dad waving from
a bridge at the edge of town. After all these miles, in a village
far from home, he is waiting for me in exactly the right place.
I can't stop smiling.

We pull into the tiny boat harbor and Pat holds the canoe
against the bank as I step out. My dad scrambles down from
the bridge, stepping carefully so as not to slip, and opens his
arms for a hug. When he squeezes me, a collection of memories from the last six months comes rushing back. His voice
on the satellite phone, answering each call with his usual

calm. The cooler of food he and my mom drove fourteen hours to deliver to us. His proud reports about the local birds he'd seen. His yodels of "Caarooooliiine!" in the Tombstone Mountains. His bright words of support as Pat and I shivered in a dark ravine. The faith that he's had in me, *no matter what.*

As he steps away, his voice sounds choked. "I don't think you know what you just did."

But, for the first time on the entire trip, I think I *do* know. It has less to do with the number of miles we traveled than with the clarity six months in the wilderness has offered. The peace that comes from letting go. The certainty that comes from being loved. I know we may never do anything quite this grand again. But I will also never forget what is possible.

I can think of only one thing to say in response. "Thank you, Dad."

When our jet takes off over Kotzebue Sound the next morning, I peer out the plane window to look for our last campsite. Pat presses his face against the scratched plastic next to mine as the engines vibrate beneath us, their roar loud and unfamiliar. My stomach lurches with the sudden speed as we hurtle through the air a hundred times faster than we can walk. Behind us, the Chukchi Sea stretches gray and limitless. Ahead, the sky holds the deep blue promise of the future. I see a smudge of white along the shoreline and imagine the swans, gathered one last time, pausing before they begin their long journey home.

Epilogue

THREE ON THE LOST COAST

Our son is barely ten weeks old when a local pilot picks us up in his blue and white Cessna at a gravel airstrip several miles from our cabin. I settle Huxley on my lap, strapping him into the front carrier as we begin to taxi. We climb above the row of jagged peaks that line Lynn Canal, the plane's engine roaring louder with the ascent, and I arrange my headset over Huxley's hood to protect his ears against the noise. Wearing oversize earphones and a microphone, he looks the part of a miniature aviator as he stares intently ahead, transfixed by the plane's shiny dials and moving gauges. Sun pours down on the ice fields, now far below us.

We planned this trip months ago, when Huxley was still in my belly, and we couldn't yet imagine all the ways in which we might change when an eight-pound, dark-eyed child slid into our world. But even then, we knew one thing. Life felt simplest when we were outdoors.

After a forty-five-minute flight, we land on a narrow, rutted stretch of sand lined by a row of heavy plastic buoys that have been arranged to mark the edges of the strip. I pull wads of tissue paper out of my ears and remove the headset from Huxley's as we roll to a stop. A field of wild strawberries greets us as we step down from the plane. Several hundred yards away, the surf breaks offshore, white tendrils of foam curling against a green-blue sea. Pat and I grin hugely as we unload our backpacks and rafts. The pilot whistles under his breath. Huxley's eyes crinkle. It's impossible not to smile at the sight of all this raw beauty.

The pilot revs his engine to turn around, then gives us a thumbs-up sign before barreling into the sky. We watch the plane become smaller, rising over the Alsek River and high above the trees before disappearing. We're alone on the Lost Coast, a remote stretch of beach that fronts the Gulf of Alaska, far from any towns or other people. This part is familiar. The fact that there's a small baby nuzzling at my breast is not. There are no manuals for backcountry travel with an infant. There's no way to prepare for becoming a parent.

The first several miles of hiking on smooth sand pass easily as Huxley sleeps on Pat's chest. We take a snack break and check the map, which shows a river running parallel to the beach. Peeking over the grass-covered dunes, we can see a picturesque, slow-moving channel and decide to take advantage of the calm water. I change Huxley's diaper beneath an alder on the riverbank while Pat inflates our boats. Huxley is smitten with the fluttering leaves, smiling as the wind rustles the natural mobile above his head. He naps again on my lap as we float, swaddled by his lifejacket. Other than the splash of our paddles, the river is silent.

That evening, we set up camp on a long gravel spit and pitch our tent among weathered white driftwood. I zip my sleeping bag together with Pat's, and wedge my feet between his calves for warmth as I often do. We've spent hundreds of other nights just like this, two of us alone in the wilderness. Tonight, of course, is different. I lean my face close to Huxley's and feel his breath, warm and moist against my cheek. He looks like a tiny polar bear, round face framed by a white fleece suit, with silly, adorable ears sewn onto the hood. Curving my body protectively around his, I tuck his hands beneath the sleeping bag and study his gently pursed lips. With darkness come the fears I held at bay while we were traveling in the sunshine. Our son is so small. This place is so big. Each rustling or splash makes me jump as I anticipate a bear lurking nearby or a wave lapping too near. When Huxley wakes to feed at 4:30 a.m., the first light filters through the tent's fabric. He fusses briefly before latching on, one hand resting against my chest. The comfort of this early-morning routine lulls me into drowsiness, and I doze off again until the sun begins to warm our sleeping bags.

When I open my eyes, Huxley is awake, gazing up. He's mesmerized by the way the breeze ripples the fabric, making shadow puppets across the tent walls. He looks over at me and smiles. *Why don't we do this more often, Mama?* Outside, the morning is stunning, fog rising from the beach in an orange haze, driftwood backlit by the sun, waves crashing offshore. I scan the spit with my binoculars and spot a large raft of scoters gathered beyond the surf, their black and white plumage gleaming against the sea. Despite my fears, there are no bears anywhere.

Each of the following days, we cover barely half the miles we normally would without a baby. I'd like to say we see twice as much, but that's only true if you include the micro-views, the odd places we find ourselves when Huxley's internal clock tells us to pause. The leeward side of a small, grassy dune; the bank of the Dangerous River; the wind shelter we construct from a collection of driftwood. While we're stopped, we do things we used to consider a waste of time. We lie down in the sand. We unroll a sleeping mat and let Huxley wiggle around on his back while we boil water for tea. We follow a set of wolf tracks as they meander down the beach, then vanish at the edge of a dune.

Everything looks different from this perspective. Mount Fairweather shimmers fifteen thousand feet above us, its glaciers bright white against the blue sky. Several summers ago, we stood on its summit, gazing down at the coast from above. Today, we sit on the beach staring up. Light dances along the stalks of the rye grass as it blows in the wind. The sand is fine and silky against my palms. As we begin to hike again, small, pale sanderlings scurry along the edge of the surf line. For the first time, I notice how their movements mirror the rhythm of the ocean. Up and down. Back and forth. In and out. Like a heartbeat, like breath. With each wave, dozens of tiny feet beat a steady pulse against the sand.

I knew that a baby would change our lives. What I hadn't realized is that this doesn't mean we must let go of what we love. Only now do I see that my worries about losing myself, or us, or our desire for adventure, were misplaced. Nearly two years later, we are still the couple who left Bellingham in a hailstorm. We are the same two people who stared out the plane window at the village of Kotzebue, bathed in gratitude for all that is wild. Now, we are also parents. We will continue to navigate by the only means we know: one stroke, one footfall, one moment at a time. Perhaps our child will also come to crave wild places someday, although for now he needs only warmth and milk and a clean diaper.

Back in Anchorage, I have returned to my work as a biologist, studying the Arctic rather than walking across it. Pat designs and constructs homes for other families. Most of the year, we live in a cozy house in downtown Anchorage that Pat built. In the summers, we disappear to our cabin on Lynn

Canal. For now, this life suits us. But we always have an eye trained on more distant horizons.

One morning, heading up the beach in search of a protected place to change Huxley's diaper, we notice a green globe shining among the driftwood. As we approach, I see that it's a glass buoy, perfectly round and larger than any I've ever seen. A traveler on this coast like us, the buoy likely arrived here from Asia, where glass fishing floats have only recently been replaced by plastic ones. I pull it from the sand and lift it up to examine. It's heavy, heavier than Huxley, but I've decided almost before I touch it that it's coming with us. Using Huxley's orange life jacket and the straps from my raft, we rig a temporary yoke to attach the buoy to the top of my pack.

Pat lifts Huxley into his arms and we return to the water's edge. The ball sits high above my head, green glass backlit by the sun. Huxley is tucked tightly against Pat's chest. Our shadows dance tall across the textured dune. Together, we continue down the beach, a family of three. The ocean crashes. The buoy glows. We walk.

ACKNOWLEDGMENTS

In stitching together the various threads of this story, I've come to appreciate the true meaning of endurance. Navigating through a sea of words and over the crevassed terrain of writing and revision has tested my abilities and often my resolve. Without a capable and dedicated crew, I would have floundered long ago. Their contributions span everything from logistics to bird identification, literary inspiration to publishing advice.

Thanks go first to my parents, Rose and Willy Van Hemert, for insisting since my childhood that we can all benefit from a little fresh air. Together, they also instilled in me a sense of family that runs steady and strong. They have done so much to support our trip and this book, including shuttling gear, organizing food drops, and providing childcare while I wrote. My sister, Ashley Van Hemert, has alternatively served as my grounding force, my voice of conscience, and my safety officer. She was also an early and eager reader. My brother, Hendrik Van Hemert, is a wizard with all things technological, and helped to narrate our journey from a distance. Joanne Farrell deserves enormous credit for trusting her children to follow their dreams, no matter where they may lead. Richard

Farrell has offered, in addition to his encouragement, astute medical advice to us in the middle of the wilderness.

There have been many champions for this book; they have provided both critical guidance and much-needed cheerleading. I am grateful to my agent, Bonnie Nadell, of Hill Nadell Literary Agency, who believed in the project from the beginning. She helped me to see that the birds would guide me, as they always had before. Austen Rachlis shepherded this book into good hands, and provided valuable comments and suggestions along the way. Huge thanks go to Tracy Behar, my editor at Little, Brown and Company. Her incisive wit, enthusiasm, and attention to detail made the editing process a joy. From her sage advice, I've learned that less is often more.

Ian Straus, Jessica Chun, Katharine Myers, Lucy Kim, and others at Little, Brown and Company have done much behind-the-scenes work to get this book to readers. I have benefited greatly from Peggy Freudenthal's thorough and careful work. David Coen provided deft and necessary copyedits.

Jill Fredston, Hannah Moderow, and Elizabeth Colen shared their time, literary insights, and, most of all, encouragement. They have been with me since the earliest days of this book, as friends, readers, and sounding boards. Andromeda Romano-Lax and Deb Vanasse provided editorial advice that pushed me to find the heart of the story. Dan Ruthrauff, shorebird enthusiast and loyal friend, helped ensure that I got my bird facts straight.

Members of my growing community of writers, in Alaska and beyond, have offered advice and inspiration. There are too many people to name, but special thanks go to Kate Harris, Ken Ilgunas, Rob Wesson, Bill Streever, Cinthia Ritchie, and the late Louise Freeman-Toole. My early love for literature was fostered and inspired by many amazing teachers in the Anchorage School District, who worked tirelessly, and often thanklessly, to share their passion and knowledge with young minds. The Rasmuson Foundation provided financial support, and gave me the confidence I needed to maintain my momentum. The Sustainable Arts Foundation similarly offered recognition and motivation at a critical time.

Colin Shanley has been endlessly generous with his time, his home, and his talents. He has hauled gear, stacked logs, and been a partner in adventure with us. The beautiful, precise map of our route is his creation. Lily Weed built a lovely website and helped me navigate through the labyrinth of social media.

Thanks to my colleagues and friends who have helped to shape my love of birds and my perspective on science. These include Colleen Handel, John Pearce, and Bill Calder (in memoriam), and the dozens of people with whom I've had the privilege of spending time in the field. Karen Loso and other dear friends dragged me outside when I needed it most.

Many individuals contributed their time and expertise to make our journey possible. These include Roman Dial, Erik LeRoy, Zach Shlosar, Colin Angus, Andrew Skurka, Sheri at Alpacka Rafts, Tom and Celest Gotchy, George Dyson, Luc Mehl, and Richard Gordon. Drake Olson has transported us safely to amazing and remote places.

The generosity of people we met along the way cannot be overstated. Sharing food, hospitality, and stories, they reminded us that our journey was about so much more than wilderness. I want to extend particular thanks to Francesco Bruti, Ricky Ashby, Katherine and Rick from Bella Bella, Sarah and Gebhardt at Tagish Wilderness Lodge, Dorothy and her family at Running River, Lee John Meyook at Herschel Island, and the communities of Fort McPherson, Aklavik, Kaktovik, Arctic Village, Anaktuvuk Pass, and Noatak.

ACKNOWLEDGMENTS

I am grateful to the individuals and organizations who share the common goal of conserving birds, wildlife, and the places they call home. Without their dedication and concern, our planet would be a much starker place. I also owe much appreciation to the birds themselves, whose very existence is a daily source of wonder.

My sons, Huxley and Dawson Farrell, help me see each bird as though it is my first. They also tolerated my absence while I wrote, and reminded me to take time to play in the mud.

Finally, my deepest thanks go to Patrick Farrell. Together, we experience more than I ever could alone. As a partner, friend, and husband, he has never wavered, no matter how steep the slope or how fast the current. As an artist, he is both gifted and generous; all of the illustrations and many of the photographs in the book are his. Without his support, the trip, the book, and our rich and surprising life would not be possible. For this and everything else, I am eternally grateful.

INDEX

INDEX

INDEX

INDEX

ABOUT THE AUTHOR

Caroline Van Hemert, Ph.D., is a biologist, writer, and adventurer whose journeys have taken her from the pack ice of the Arctic Ocean to the swamps of the Okavango Delta. She currently works at the US Geological Survey (USGS) Alaska Science Center and regularly publishes scientific articles about birds and other wildlife in the north. Her research and her expeditions have been featured by the *New York Times,* MSNBC, *National Geographic,* and more. She lives in Alaska with her husband and two young sons.